To acknowledge the wonder

The title is derived from a sentence of Einstein:

'The nice thing is that we must be content with the acknowledgement of the wonder . . .'

This is contained in a letter of Einstein written in 1952, a copy of which is in the Archive at the Princeton Institute for Advanced Study. Part of this letter is quoted by G Holton in *Some Strangeness in the Proportion*, ed Woolf (New York: Addison-Wesley 1981).

To acknowledge the wonder

The story of fundamental physics

Euan Squires
Department of Mathematical Sciences,
University of Durham

Adam Hilger Ltd
Bristol and Boston

British Library Cataloguing in Publication Data

Squires, E. J.
To acknowledge the wonder: the story of fundamental physics.
1. Nuclear physics
I. Title
539.7 QC776

ISBN 0-85274-786-1
ISBN 0-85274-798-5 Pbk

First published 1985
Reprinted 1986

Consultant Editor: **Professor R H Dalitz,** Department of Theoretical Physics, University of Oxford

Published by Adam Hilger Ltd
Techno House, Redcliffe Way, Bristol BS1 6NX, England
PO Box 230, Accord, MA 02018, USA

Typeset by Mid-County Press, London SW15
Printed in Great Britain by J W Arrowsmith Ltd, Bristol

Contents

Preface

Fundamental physics is about the things of which the observed world is made and about the laws which govern their behaviour. Implicit within it is the belief that there exist reasonably simple constituents, obeying reasonably simple laws, and that these provide an explanation of all 'physical' phenomena. Which phenomena are to be regarded as physical, in this sense, depends upon the success of physics; it is certainly a large class, and it grows with time.

This book attempts to describe the present situation in fundamental physics and, in particular, the recent spectacular developments. We enter the world of atoms and their nuclei, the protons and neutrons of which these are made and which in turn contain, forever trapped within them, the mysterious quarks; we find out about space and time and about the strange, almost magical, dualism of quantum theory; we learn how particles interact and how these interactions can all be derived from beautiful general principles; we try to understand the relations between quarks, electrons and other, similar yet excitingly different, particles that we meet on the way; we also discover some facts about our Universe and its early history and even see how its present properties can be understood in terms of the things we have learned. We shall discover that although much is known, there remain many questions to which the answers are, as yet, unknown.

I am grateful to many people who have offered advice and suggestions which have, I am sure, made this book more appropriate for its purpose than it would otherwise have been. In particular, I must mention Alexandre d'Adda, who read the first draft; Timothy and Clare Squires who commented at an early stage; John Rowell, Head of the Mathematics Department at Theale Green School, Reading, and several of his sixth form pupils; also Professor R H Dalitz for many constructive and detailed comments and Professor G H A Cole for helpful advice. My thanks are also due to Maurice

Jacob, for hospitality in the CERN Theory Division, where this work was started; to Jim Revill, of Adam Hilger Ltd, who encouraged and assisted with the project from the beginning; and finally to Mrs Sheila Cottle, who has typed a seemingly endless series of revisions.

E J Squires

Chapter One

in which we discover where we are going, and why,
we meet the atom and the electron,
we learn about electric and magnetic forces,
and how these are related to light,
we enter the amazing and beautiful world of quantum mechanics,
and come to understand the structure of atoms,
and thereby most of observable physics.

§ 1 The story of physics

Just outside Geneva a tunnel is being built. With its associated apparatus it will cost about £300 million†. It is called LEP; it does not go under the Alps, it will carry neither cars, nor trains, nor people, it will not provide access to oil, or minerals, or precious stones. When it is ready, tiny particles will travel the 27 kilometres of its circular path at a rate of about 10 000 ($\equiv 10^4$) times/second. Occasionally they will collide—and it is to observe what happens in these collisions that the tunnel is being built. The collisions are expected to produce events that happen nowhere else in our Universe, and, indeed, have not happened before, except possibly when the Universe was a tiny fraction of a second old.

An hour's drive along the motorway from Geneva and 3000 metres below the summit of Mont Blanc, this time in a tunnel that was built to carry cars through the Alps, 160 tonnes of detecting equipment have been installed in order to discover whether particular particles might occasionally change spontaneously into other particles. The scientists responsible for the experiment know that an individual particle will most likely take far longer than the age of the Universe for it to perform this act of self-destruction (in fact, longer than 10^{30} years—whereas the Universe is about 10^{10} years old), but if they have enough of them, and if they wait long enough, they might see something happen.

A few hundred kilometres further north, in a laboratory near Hamburg, scientists are measuring a particular quantity and finding a value of $3\frac{2}{3}$. They are progressively changing the experimental conditions in the expectation that eventually the value will jump to 5. When it does, they will announce that they have confirmed the existence of 'top'.

† We shall often meet large numbers, so we shall need a notation. We will denote 1 by 1, 10 by 10, but 100 by 10×10 or, easier, by 10^2. Then $1000 = 10^3$, 1 million $= 10^6$ etc. Thus the cost is £3×10^8.

These are three of the many experiments taking place around the world in the science of *elementary particle physics*; a science on which the world spends about £7×10^8 per year. In most cases those who will use the apparatus have some vague ideas of what might happen, but they want to be *sure*, they want to see exactly what happens and, of course, they would like to see some surprises. Why?

Why do they want to know? How are they able to have any idea of what might happen? What are the reasons for wanting to know the details? Why will the surprises, when they come, cause hundreds of people to rush to their desks, blackboards or old filing cabinets to try new, or resurrect old, ideas?

To understand the answers to these questions we need to know a story. It is a story that goes far back into human history, but whose chapters have been written very rapidly in recent decades. It is a story that we can tell because, within the human mind, there is an insatiable desire to understand the Universe in which we live, to respond to each new experience with *why?* and to seek explanations for life's varied phenomena. It is a story that is undoubtedly a great tribute to the power of the human intellect but is also, of course, a far greater tribute to the one (call it God, or chance, or whatever) who conceived its subject. It is a story of which we certainly do not know the end, but which is, already, a great, indeed a thrilling, story. It is the story we try to tell in this book.

§2 A story worth telling

But why bother? Is there any reason, apart from the satisfaction we all derive from telling a good story, why I should, at this time, wish to write this book? About five kilometres from the LEP tunnel mentioned previously is Geneva airport, at which people arrive from all over the world to participate in the varied international organisations that have their headquarters in Geneva. They come to talk education, health, culture, peace, etc, and if asked the purpose of their visit they will refer to one of these activities. Usually, their reply will be understood. Rightly or wrongly, we all think we have some reasonable ideas about what the World Health Organisation does, about the purpose of the General Agreement on Tariffs and Trade, etc. It is very different, however, if they come to work at CERN, the European Laboratory for Particle Physics. Whatever message would be conveyed to the average person by the fact that a physicist had come to CERN to look for gluon jets would almost certainly be false!

To tell someone, newly met at a party for example, that one studies elementary particle physics is almost certain to stop the conversation—and perhaps even the acquaintance!—immediately. If the listener is bold he will ask 'what's that?' 'Well, I guess, it's about what everything is made of.' (Roughly true, I suppose). 'Oh, yes, atoms and things. I remember we did that at school—I found it all a bit boring.' 'I suppose what you do has something to do with these nuclear power stations and such like.' He is, maybe, too polite to suggest that he really thinks you must be designing bombs. Perhaps, when this book is completed, the questioner can be recommended to buy a copy and, at his leisure, discover what it is all about.

More seriously, many of us who are considered as experts in the subject have been asked by philosophers, by theologians and by others who really desire answers, what *is* known? What progress has been made? Does what is known have any relevance to what *I* do or think about? Pure mathematicians, in particular, are totally sceptical about our attempts to explain, and cannot share our confidence in the things about which we are reasonably certain; they are thus inhibited in offering their assistance to solve the many problems that still remain.

It is, then, a story that needs to be told. What is more, this is a good time at which to tell it. Recent years have seen much progress; particular ideas have been confirmed (and others quietly forgotten), so that there is now a 'standard model' which, whilst almost certainly incomplete, is, with equal certainty, unlikely to be wrong. The story is no longer a catalogue of unrelated facts; rather it is about several elegant, simple and unifying themes which together provide an understanding of essentially the whole of physics.

I have endeavoured to make this story accessible to a wide variety of readers, so, at the risk of boring those already well acquainted with the material of the earlier sections, who may nevertheless find the point of view somewhat novel, I have assumed very little previous knowledge of physics. Some readers, however, might be overawed by the appearance of 'mathematics' in the book. The purpose of mathematics is to make difficult things easier and, in several cases, I have not been able to proceed without its aid. Familiarity with mathematical notation and manipulation (at about the standard of A level students of mathematics or physics) will therefore be required by readers who wish to follow all the details. In spite of this, I hope that readers without such familiarity will find that there is sufficient non-mathematical description of what is happening for them to follow the essential features of the plot. The degree of difficulty varies throughout the book—some ideas just are more subtle than others; where the going becomes hard, I can offer the encouragement that it will probably become easier again.

It is perhaps not surprising that the logical way of telling our story follows approximately the historical sequence, with a few exceptions. History, however, is not our primary concern, so we need not dwell on the mistakes and blind alleys of the historical development, nor will we be zealous in allocating credit for various major steps.

It is worth saying here that this is not intended as a text book. There are no questions at the end of the sections, and I hope nobody tries to set an examination on the material in it. I trust this is not incompatible with the additional hope that schools, and other educational institutions, might find it useful.

I dedicate the book to the vast majority who *don't* know, most of whom will never know and, I imagine, some at least of whom don't care, but who have in many cases contributed financially to this story and enabled some of us to enjoy its discovery.

§3 Our comprehensible world

We have all forgotten the time when we first learned about 'cause and effect'. Soon after our introduction to the world, we discovered that the things that happened around us were in some way related: particular actions (usually) gave rise to particular consequences. Thus we became familiar with the possibility of 'explaining' things; Mother comes *because* I cry, a toy moves *because* I push it, a ball drops because I lift it and then release it, etc. Soon we expect to explain everything, and come to realise that each explanation requires others—*why* does Mother come when I call, *why* does a released object fall?

The study of fundamental physics is a continuation of this process. The physicist retains a simple faith that phenomena *can* be explained; that things don't 'just happen'; that the question 'why?' has an answer and that the new questions 'why?' suggested by that answer will also have answers. It seems, so far, that this faith has been justified. Should we be surprised by this? Einstein once wrote†: 'the eternal incomprehensible thing about the world is its comprehensibility.' If this amazed Einstein then it should certainly amaze us. What is more, the *possibility* of understanding the world is surely an encouragement to try.

In order to ask, and certainly to answer, the questions, it is necessary to examine very carefully the phenomena we seek to explain. We need, first, to ask 'what?' Again, this is a simple continuation of the behaviour of a child, who continually 'investigates'—in order to clarify answers and to suggest new questions. Thus, physics is an *experimental* science, a fact which we must emphasise here since the subsequent discussion will be largely *theoretical*, in that we shall not, in general, concern ourselves with the experimental methods used to determine particular facts. (This would make a separate, exciting story which would be a great tribute to experimental physicists and

† Article in *Journal of the Franklin Institute* **221** 317 (1936).

those who support them.) It is, of course, possible that, if we were sufficiently clever, we could deduce everything—or almost everything—from 'pure thought', i.e. (almost) all phenomena would be predictable. Certainly, as we shall see, there are cases where a large variety of at first unrelated experimental facts are direct consequences of a few simple rules. (The best example here is that electromagnetism—which is the key to understanding most of observable physics and chemistry—could have been deduced from the requirement that a particular number should be allowed to vary with position and time. I hope that this statement will mean something to the reader when he or she has read §21.) We shall return later to this question of whether our world is unique, whether it is the 'only', or the 'best', etc.

A crucial aspect of any useful experiment is that it should be 'repeatable'. In order to establish that one event *causes* another, i.e. that they are 'causally' related, it is not sufficient to notice that one follows the other on one occasion; we have to convince ourselves that this always happens. This rather obvious truth is often ignored; if there is a bad summer we will be offered 'explanations' based on some other happening that may well be totally unrelated; primitive societies would seek similar spurious explanations of outbreaks of disease, etc. Of course, the statement, 'we have to convince ourselves', is to be seen in the light of the fact that we can never be absolutely sure. Clearly we can only do an experiment a finite number of times—maybe one day, October 17, 1996, for example, apples will not fall to the ground when they become detached from branches, but will stop 17 centimetres from the Earth, or even change into oranges on impact. These things are unlikely—so we ignore them—but we cannot be *sure* they will not happen.

Related to this is the fact that we never *exactly* repeat an experiment. Suppose, for example, I wish to measure the time taken for an object to fall one metre. I do the experiment and then, one hour later, I repeat it. However, the atmospheric pressure will have changed, so will the temperature, the Moon will be in a different place relative to the object, etc. Of course, these things make little difference to the result, so the important thing about repeating an experiment is that we must keep all *relevant* conditions the same. What *is* relevant is itself determined essentially by experience (i.e. by experiment coupled with theoretical extrapolation). It also depends on the accuracy to which we are working. It is certain that if we were concerned with *exact* measurement, then no experiment would ever be repeatable. In the example above, the atmospheric pressure, the temperature and the position of the Moon would have some effect on the measured time. To be content with limited accuracy is indeed often a crucial aspect of making progress towards understanding. We are probably aware that the Earth follows an ellipse in its path around the Sun. It doesn't, of course, because of a great variety of small effects, but it is fortunate that it *very nearly* does, otherwise maybe Newton would not have developed his theory of gravitation which predicted elliptical orbits for the planets.

§4 The elements

In order to determine the nature of the phenomena to be understood, and to enable experiments to be repeated, it is desirable that we should consider, first, *simple* systems. It is clear that the behaviour of some systems can be understood, at least partially, by the properties of simpler systems that are put together to make them, i.e. of their *constituents.* Thus we can conveniently think of a house in terms of its heating system, its plumbing, its bricks, etc. We are thus naturally led to ask questions like: what are the simplest constituents from which particular objects are made? What is the minimum set of constituents from which everything can be made? Much of our story involves giving answers to questions of this nature.

Familiar objects are often made out of particular substances: nails out of iron, chairs out of wood, etc, and we shall need to consider the substance, rather than the particular shape in which we find it. Thus we have a large variety of substances; some are solids (iron, wood, . . .), some are liquids (water, whisky, . . .), some are gases (air, hydrogen, . . .). This distinction, however, is not significant for our purpose (it is, in any case, dependent on temperature). We are concerned with a more important classification. There are *mixtures*, e.g. soil, cake, whisky, in which many of the properties of the constituents are retained and in which the proportions of the constituents are variable. There are *compounds*, which often show no resemblance to their constituents and where the constituents always have identical fixed proportions. An example is common salt, which is a compound of sodium, a metal which ignites spontaneously in air, and of chlorine, a very poisonous gas. Whereas the proportions of the ingredients in cake, for example, are variable this is not true for salt, in which the relative amounts of sodium and chlorine are always the same.

Finally, there are *elements*: hydrogen, oxygen, sodium, chlorine, etc, which are the basic substances of which the others are made. It is the elements we must study—when we understand them then we should be able to deduce the properties of compounds and mixtures. Before continuing with elements, however, we shall have a small digression.

§5 Space

Since we observe events at particular points of space we need to say a little about this space and its description. However, readers who find this discussion too technical and difficult will lose little by omitting it and proceeding with §6.

First, consider a straight line. This is a 'one-dimensional' object, by which we mean that, having fixed a starting point on the line, which we denote by O, the 'origin', and a direction along the line, then any other point P can be uniquely described by giving *one* number (this is the 'one' of 'one-dimensional'), namely the distance OP, which will be positive or negative according to the initially chosen direction. In figure 5.1 we have a point P, specified by OP = 2 cm, and a point Q with OQ = −1 cm.

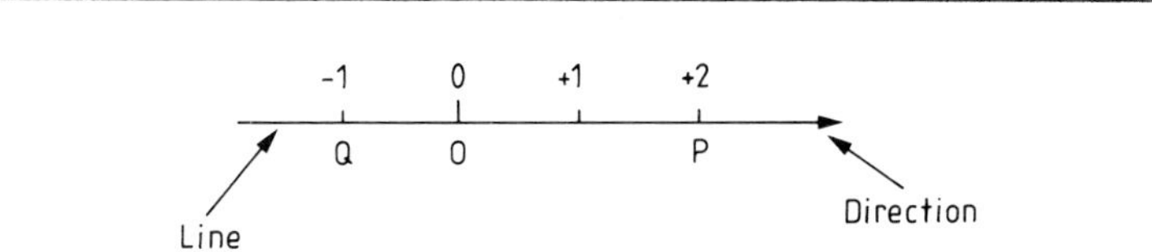

Figure 5.1 A space of one dimension. Points, e.g. P and Q, are specified by a single number, the distance from O measured in cm. This number is positive if the point is to the right of O and negative if it is to the left.

Now consider a *plane*, e.g. the page you are reading. This is a 'two-dimensional' object. To specify a point P we need an origin, O, and two lines in the plane, which for convenience we take at right angles (i.e. 'perpendicular' to each other), each with an associated direction. An example is shown in figure 5.2. A point P is now uniquely described by *two* numbers, namely, the distances along each of the

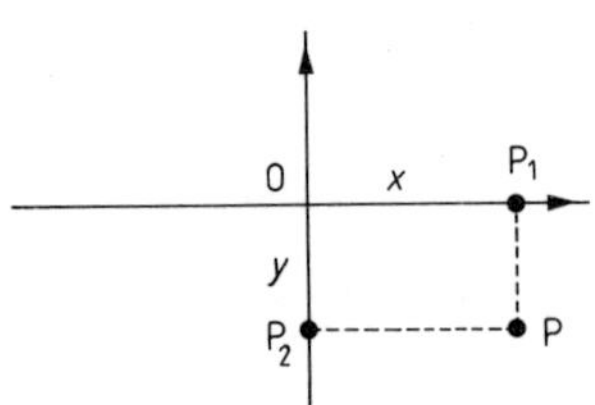

Figure 5.2 A space of two dimensions, in which two numbers, namely the lengths OP_1 and OP_2, are required to fix the position of a point P.

two lines, e.g. OP_1 and OP_2 in the figure. It is often convenient to call the two original lines the x axis and y axis respectively, and to call $OP_1 = x$ and $OP_2 = y$. For the point P in figure 5.2, x is positive and y is negative.

The two-dimensional description of a point shows a feature which does not exist in one dimension. Namely, there is freedom in the choice of the directions for the two perpendicular lines, i.e. the two axes. For example, in figure 5.3 we see a different choice from that in figure 5.2. The point P is the same, but the analogues of P_1 and P_2, which we now call P_1' and P_2', are different. Thus, the numbers with which we describe P, namely $x' = OP_1'$ and $y' = OP_2'$, are not the same as x and y, although P is the same point.

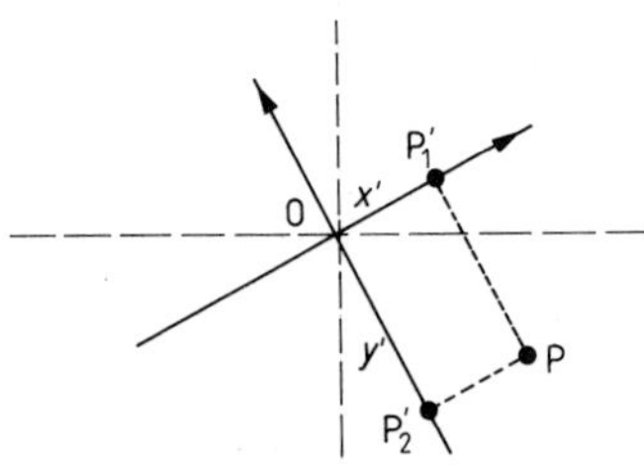

Figure 5.3 The same space as in figure 5.2 but with different axes. Here the same point P is specified by different numbers OP_1' and OP_2'.

Many of our readers will be familiar with Pythagoras' theorem†, which will tell us that the distance OP can be expressed in terms of x, y or x', y' by

$$(\mathrm{OP})^2 = x^2 + y^2 \qquad \text{from figure 5.2} \tag{5.1}$$

or

$$(\mathrm{OP})^2 = (x')^2 + (y')^2 \qquad \text{from figure 5.3.} \tag{5.2}$$

Thus, although the numbers x, y vary according to the way the axes are chosen, the combination $x^2 + y^2$ does not. This is a mathematical way of stating the obvious fact that the distance OP does not vary when I merely rotate the axes used to describe the point P. We can make it sound very profound if we express it by saying that $x^2 + y^2$ is *invariant* under a rotation of the axes.

Finally, we allow the point P to move away from the plane of the page. Clearly, then, we need a further number, i.e. the distance from P to the plane, to describe the point uniquely. Thus space has *three* dimensions: given an origin, O, and three *mutually perpendicular* lines we can specify any point by giving the distances along the three lines. These distances are often denoted by x, y and z. (A good picture of this is to take two adjacent walls and the floor of a room. They all meet at a corner—which is our origin O. Then a point P can be defined uniquely by giving its distances from the two walls and from the floor.)

It is clear that in the full three-dimensional space we again have freedom in the choice of our three axes, so that for a fixed point P, we can obtain different numbers x, y and z. However, as before the distance OP^2 is always given by

$$(\mathrm{OP})^2 = x^2 + y^2 + z^2 \tag{5.3}$$

which quantity is therefore invariant under any rotation of the axes.

In the above discussion the distance between two points, e.g. O and P, in space has been denoted by OP. We shall sometimes use instead a single letter, e.g. r (just as we used x for OP_1, etc). We often want to describe not only the length of the line, but also its *direction*. In that

† A number multiplied by itself is said to be 'squared', and if x is the number, then x squared is written as x^2. Pythagoras' theorem says that in any triangle containing a right angle the squared length of the side opposite the right angle is equal to the sum of the squared lengths of the other two sides. (Why not draw such a triangle and check that it works?)

case we shall use bold face type, i.e. **OP** or $\boldsymbol{r}$. Thus, whereas $s = 2r$ means a distance twice the distance r, the equation $\boldsymbol{s} = 2\boldsymbol{r}$ defines a line which is twice as long as $\boldsymbol{r}$ and in the same direction. Quantities which have direction as well as length are called 'vectors'.

In figure 5.2, the vector **OP** is the line from O to P. The distances OP_1 and OP_2 are called the 'components' of **OP** in the directions of the two axes.

Before we return to the elements it is convenient to introduce here the idea of 'units', i.e. the quantities in which we make measurements. To say that a distance is 3 is a meaningless statement: 3 what? A distance must be given as a multiple of a certain *unit distance*. Of course we are free to choose our unit, and different choices are convenient for particular situations. The metre (m) is now the generally accepted basic unit of length, though metres times various multiples of ten are sometimes more suitable, as, for example, in our specification of the points P and Q in figure 5.1, where we used the centimetre ($1\ \text{cm} \equiv 10^{-2}\ \text{m}$)†.

Other types of unit are, of course, required. We cannot measure a *time* in space units, so we need a unit of time. We shall normally use the second (s). Similarly we need a unit of mass; the accepted choice here is the kilogram (kg). We shall also need a unit, degrees centigrade, in which to measure temperature. Finally, in §7, we shall meet the idea of electric charge. Whether or not we introduce a new type of unit for this is a matter of convention. This issue and others related to units are discussed in the appendix.

Combinations of these units can be used for other quantities. For example, the *velocity* is the distance moved in a unit of time; thus it has the units‡ of (length) × (time)$^{-1}$ and can be measured in m s^{-1}. If we have an object at position $\boldsymbol{r}_1$ at a time t_1, and at $\boldsymbol{r}_2$ at a later time t_2, then its velocity is the distance moved, $\boldsymbol{r}_2 - \boldsymbol{r}_1$, divided by the time taken, $t_2 - t_1$. Thus:

$$\boldsymbol{v} = \frac{\boldsymbol{r}_2 - \boldsymbol{r}_1}{t_2 - t_1}. \tag{5.4}$$

† Recall the footnote on p 3. Now we are going to small numbers: one tenth $= 1/10 = 10^{-1}$, one-hundredth $= 10^{-2}$, etc.

‡ We use the word 'unit' to describe the type of unit (e.g. length, or mass, etc) as well as the magnitude. The word 'dimension' is often preferred for the first usage (see appendix), but this can cause confusion.

As the notation implies, velocity, being the difference of two vectors, is itself a vector. Strictly speaking $\boldsymbol{v}$, given by equation (5.4), is the average velocity during the time interval from t_1 to t_2. To get the actual velocity at t_1, it is necessary to make this interval very small. Readers who know calculus will then replace (5.4) by $\boldsymbol{v}=\mathrm{d}\boldsymbol{r}/\mathrm{d}t$.

We shall also meet the acceleration of an object. This is the rate at which the velocity changes. Thus, given a velocity $\boldsymbol{v}_1$ at t_1, and $\boldsymbol{v}_2$ at t_2, the acceleration is defined by

$$\boldsymbol{a}=\frac{\boldsymbol{v}_2-\boldsymbol{v}_1}{t_2-t_1} \tag{5.5}$$

or, again for readers who know calculus, $\boldsymbol{a}=\mathrm{d}^2\boldsymbol{r}/\mathrm{d}t^2$. The units of acceleration are clearly $\mathrm{m\,s^{-2}}$.

§6 Atoms

In our pursuit of simplicity, we are considering the *elements*, of which there are 92—with widely varying properties: hydrogen is a highly inflammable gas, helium is a gas which is not inflammable and indeed is almost completely inactive chemically, carbon is an element that appears in many forms (e.g. diamond, charcoal), iron is a familiar metal, etc.

At first sight, these elements are not characterised by a 'size'; we can have iron, for example, in all sizes and shapes. It is natural to ask, however, whether we can have an arbitrarily small piece of iron; can we continue to subdivide it into ever smaller pieces and still have a substance that is iron? This question of complete *homogeneity* or *discreteness* was a long discussion in science, which was finally ended when it was realised that all elements were made up of particles, which we call 'atoms'; e.g. hydrogen is made of hydrogen atoms, iron of iron atoms, etc. That this fact is not obvious to our everyday experience is accounted for by the fact that atoms are very small, about 10^{-10} m.

These 92 kinds of atoms are 'building blocks' of all matter. If we are given the properties of the 92 atoms (i.e. their masses and the forces between them), then, in principle, we can calculate the properties of all other substances. Of course, the 'properties' assigned to each atom would be very complicated; they would have to explain, for example, the widely different behaviours of the various elements and of the compounds which are formed from them. Thus a lot of outside information would have to be given. Fundamental physics seeks to eliminate the need for outside information; we want as much as possible to be *calculable* and as little as possible to be *given.* The glorious success of this venture, with regard to atoms, is the story of this chapter, in which we shall see that *all* the known properties of atoms can be deduced from very little input. It is a success which would have been incomprehensible to scientists around the start of this century; their knowledge was inadequate for them to contemplate such a venture; to them atoms were outside the range of things that the human mind could reasonably hope to understand.

Before ending this section we should for completeness briefly mention the evidence on which the atomic theory of matter was first based (of course the possibility goes far back into scientific thought). The first real indication comes from the observation noted above that compounds contain fixed proportions of their constituent elements. A natural understanding of this would come if a compound was made of objects (which we call molecules) *each* of which consists of N_1 atoms of the first constituent, N_2 atoms of the second, etc, where $N_1, N_2, \ldots$ are fixed integers. Some examples are:

salt molecule = 1 atom of sodium + 1 atom of chlorine
water molecule = 2 atoms of hydrogen + 1 atom of oxygen.

More quantitative evidence came from the 'kinetic theory' of gases (or, in general, the science of 'statistical mechanics'). A gas consists of atoms or molecules, moving almost freely, with occasional 'collisions'. The 'average velocity'† of these atoms determines the temperature (high temperature ↔ high velocity). It is then possible to show that, for a fixed amount of gas, the pressure (P) multiplied by the volume (V) and divided by the temperature (T) is approximately a constant, even though individually these quantities may vary. We write this relation as

$$PV/T = \text{constant}. \tag{6.1}$$

Note that the pressure is the force required per unit area to keep the gas confined in the volume V. (We can easily understand equation (6.1) in a loose sort of way. If we increase the volume, keeping the temperature fixed, then the atoms will hit the sides less frequently—thus reducing the pressure. Also, if we increase the temperature, at fixed volume, then the pressure will increase because the atoms striking the sides will be moving faster.) This equation is experimentally verified.

We must now, as promised, turn to our explanation of the properties of atoms. To do this we must first be introduced to one of the most beautiful of all physical phenomena—we shall meet the electric force.

† Since velocity has a direction it is a vector. For a gas in a stationary closed container, its average will be zero since it is equally likely to be in any direction. Thus we take the average of the magnitudes or, more easily, of the squares of the magnitudes.

§ 7 The electric force

We have so far discussed the various possible constituents of matter. But constituents have to be held together in some way; objects 'know about' each other, i.e. they *interact*. All this is the subject of forces, although, as we shall see later, forces and particles are very closely related.

A force is *defined* as something that gives an object an acceleration, i.e. that changes its velocity. More precisely, the force on an object is its acceleration multiplied by its mass:

$$\boldsymbol{F} = m\boldsymbol{a}. \tag{7.1}$$

This equation, usually called Newton's second law of motion, implies that, for a given force, the effect is smaller the larger the mass; hence mass is sometimes called 'inertial' mass—it resists changes in velocity. We can also see from this equation, and the end of §5, that the units of force are (mass) × (length) × $(\text{time})^{-2}$, and that it can, for example, be measured in kg m s^{-2}.

The first force we must discuss is the one responsible for almost all readily observable phenomena, namely, the electric force. As we shall try to explain, this is about the simplest possible force that could exist between two small objects. We shall describe it in some detail, although these details are not essential for following our story.

Associated with any object is an 'electric charge'. This is simply a number—either positive, negative or zero. Suppose we have two objects, denoted by 1 and 2, which we here take to be points (i.e. to have zero size), with electric charges Q_1 and Q_2 respectively. Then each object exerts a force on the other of magnitude given by the product of the two charges divided by the square of the distance between them. Thus we have Coulomb's law:

$$F = \alpha Q_1 Q_2 / r^2 \tag{7.2}$$

where r is the distance between the two objects and α (alpha) is a constant whose value depends on the units in which the charge is measured†.

Equation (7.2) is an experimental law, which we could verify by measuring the acceleration of one of the objects. Of course, in an experiment we could not use zero-size objects, and we would be content with small objects. However what does 'small' mean? Size, i.e. length, has a unit and the actual number associated with the size would depend on the unit used. This is a situation we shall often meet and it is necessary to define small in terms that are independent of the units. This means that we must take a ratio of two lengths. It is reasonably obvious that the relevant ratio in this case is the size of the object divided by their separation r (there *are* no other lengths in our problem); this ratio should be very small compared to one.

As we saw in §5, acceleration is a vector (the particles would begin to move in a particular direction), hence so is force (equation (7.1)). Thus equation (7.2) is not complete, since it does not tell us the direction of the force. Experiments show that the force is along the line joining the two particles, i.e. along the vector $\boldsymbol{r}$. Then the force that 1 exerts on 2 is given by rewriting equation (7.2) as

$$\boldsymbol{F}=\frac{\alpha Q_1 Q_2}{r^2}\,\hat{\boldsymbol{r}} \qquad (7.3)$$

where $\hat{\boldsymbol{r}}$ is defined to be of length one (hence the 'hat') and in the direction of the line joining 1 to 2 (see figure 7.1). We use, here and elsewhere, r for the length of the vector $\boldsymbol{r}$.

There are several features of equation (7.3) that we should note:

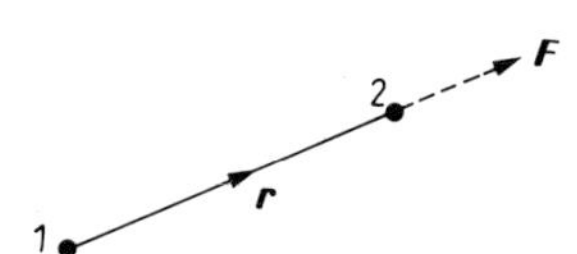

Figure 7.1 Showing the direction of the force between two point charges denoted by 1 and 2.

† Many readers will be familiar with SI units in which the charge is measured in coulombs (C) and the constant α is written as $(1/4\pi\varepsilon_0)$. See appendix for further discussion.

(a) We did not really need an experiment to tell us that $\boldsymbol{F}$ would be along $\boldsymbol{r}$. The argument is simple: what other direction could possibly be involved? Recall that we are thinking of a situation with only our two structureless particles (all other effects being assumed too weak to affect our measurements), so the only line† that exists in our world is $\boldsymbol{r}$. Suppose, for example, that we tried to say that the force was at 45° to this line. This would clearly define a cone (see figure 7.2) and we would have no way of specifying a particular line in that cone. As far as our problem is concerned, all would be equivalent. The only uniquely defined line is $\boldsymbol{r}$, so $\boldsymbol{F}$ must be along $\boldsymbol{r}$. Note that the crucial assumption here is that the Universe, i.e. space, has no preferred directions—apart from those defined by objects in space.

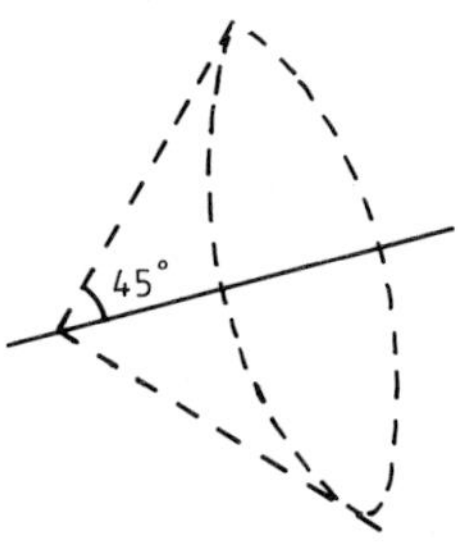

Figure 7.2 Showing why the force between two point charges must be along the line joining them. Other directions cannot be defined uniquely.

(b) We did, of course, need to specify a sign in equation (7.3). At this stage there are no theoretical reasons why it should be positive or negative and the positive sign, with a positive value for α, is the result of experiment. A consequence of this is that two objects with the same sign of charge, in particular two identical objects ($Q_1 = Q_2$), repel each other.

(c) The magnitude of the force in equation (7.3) decreases with distance like r^{-2}. This decrease is essentially a 'geometrical' effect. (To

† We assume here, correctly, that the force exists even when the particles are at rest, otherwise the velocities would provide a direction, as we shall see later (§10).

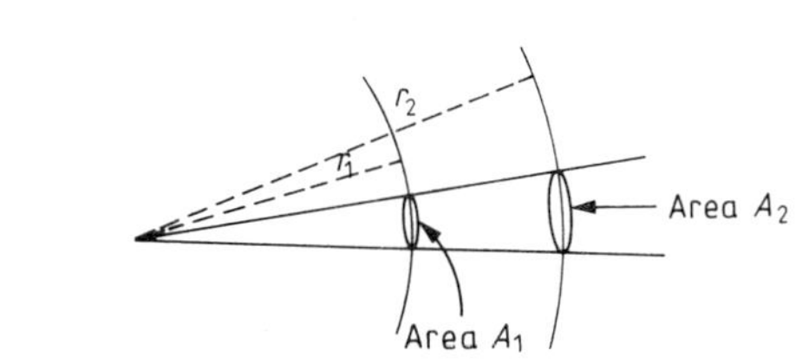

Figure 7.3 Showing the areas defined by the intersection of a fixed cone with spheres of radii r_1 and r_2 respectively.

see this, imagine concentric spheres of radius r about object 1, and take a fixed cone intersecting these spheres, as shown in figure 7.3. The areas of the spheres cut out by the cone (A) are proportional to the radii (r) squared, i.e.

$$A_1/A_2=(r_1/r_2)^2. \tag{7.4}$$

Thus, in order to subtend the same opening angle at the centre, an area has to grow like r^2. Conversely, for a constant area, the opening angle, and hence the apparent size, has to decrease like r^{-2}.) Later, we shall meet forces that fall off with distance much more rapidly. A force that only has the geometric factor is called a force of 'infinite' range.

(*d*) In any equation the powers of the various units must be identical on both sides—otherwise the truth of the equation would depend on which particular set of units we choose. Since we already know the unit of force (given above) we can deduce from equation (7.2) the units of α to be (mass) $\times$ (length)^3 $\times$ (time)$^{-2}$ $\times$ (charge)$^{-2}$.

(*e*) It is not evident from equation (7.3), but turns out to be true, that if we put together two objects of charge Q_1 and Q_2, then the charge on the new object would be (Q_1+Q_2), i.e. *electric charge is conserved.* Later (§21) we shall see that this fact is intimately related to the fact that the force has infinite range.

(*f*) We note, and will discuss further in §32, that the force of gravity has exactly the same form as equation (7.3) except that, instead of Q_1 and Q_2, we have m_1 and m_2, the *masses* of the objects, which are always positive, and the expression for the force has to be multiplied by a minus sign, i.e. it is *always* attractive.

To complete this section we make a trivial modification which at this stage appears totally unmotivated. We shall replace equation (7.3) by *two* equations. To do this we first introduce an *electric field*, $\boldsymbol{E}$, which has a magnitude and direction at each point of space†. Then we express the force on a charge Q placed at a point where the electric field is $\boldsymbol{E}$ as

$$\boldsymbol{F}=Q\boldsymbol{E}. \tag{7.5}$$

Secondly, we write for the electric field at a distance $\boldsymbol{r}$ from a point charge Q,

$$\boldsymbol{E}=\frac{\alpha Q}{r^2}\hat{\boldsymbol{r}}. \tag{7.6}$$

It is clear that if we use equation (7.6) for a charge Q_1, and equation (7.5) for a charge Q_2, then we obtain equation (7.2) as required.

Why have we done this? Readers may complain that the going is hard enough without introducing redundant quantities. Later we shall see that it is much easier to discuss just the *fields*, without concerning ourselves with the charges that cause them. Also fields are important when we come to think about quantum mechanics.

† It is a 'vector field', where the 'vector' implies that it has a direction and the 'field' implies that, in general, it varies with position.

§ 8 The electron

It is probably fair to date the modern study of elementary particles from J J Thomson's discovery, in 1897, of electrons. It was noticed that these particles could be drawn out of matter by strong electric fields. Because they came from the negative plate of a battery (the 'cathode') they were originally called cathode rays and they are, by convention, regarded as having negative electric charge.

By bending a beam of electrons in a magnetic field (see §10) it was possible to measure the ratio of their charge to their mass (Q_e/m_e) and later the charge was measured directly. These quantities are now known very accurately:

$$m_e = 9.109\,534 \times 10^{-31}\ \text{kg} \tag{8.1}$$

and

$$Q_e = -1.602\,19 \times 10^{-19}\ \text{C}. \tag{8.2}$$

For our purposes it will be more useful to use as a measure of the electron charge the quantity e defined by $e^2 = \alpha Q_e^2$. This has the value

$$e = 4.803\,24 \times 10^{-10}\ (\text{g cm}^3\ \text{s}^{-2})^{1/2}$$

or (8.3)

$$e = 1.518\,92 \times 10^{-14}\ (\text{kg m}^3\ \text{s}^{-2})^{1/2}.$$

The electrons are also now known to be very tiny, with a radius that is certainly less than 10^{-20} m. Note that this is 10^{-10} times the size of an atom. (Readers who wonder how we 'measure' such distances will, we hope, be eager to read on to where their curiosity will be satisfied.)

These electrons are seen to come from matter, so it is natural to assume that they are contained in atoms. Thus we have a simple model of an atom; it contains Z electrons, where Z is an integer running from 1 to 92, which characterises the particular element being considered, together with a positively charged object, with charge Z times the magnitude of the electron charge. The total charge on the atom is therefore zero. This picture fits well with the fact that it is possible to make charged atoms ('ions') which correspond to the atom with one of its electrons removed. Measurement of the charge-to-mass ratio of such ions then leads to a knowledge of the atomic masses. These are several thousand times the mass of the electrons—so we discover immediately that most of the mass of an atom is associated with its positive charge. Our picture of an atom at this stage is probably like a 'currant bun'—the mass and size mainly residing in the (positively charged) 'bun', with the electrons being the 'currants'. The truth is very different, as we shall see. For the moment, we merely note that this picture does not permit us to calculate any of the *properties* of the atom, and is therefore not very satisfactory.

Before we close this section on our first elementary particle, we introduce an important symmetry principle: to every type of particle there corresponds a particle of the same mass and opposite electric charge (and also opposite values for other quantities, similar to charge, that we shall meet later). The theoretical understanding of this symmetry will be discussed in §18. In our present context, it means that, in addition to electrons of mass m_e and charge $-e$, there can exist also *positrons* of mass m_e and charge $+e$. Such particles were first proposed in 1928 by P A M Dirac, who realised they were required in a consistent theory of the electron.

The positron is often referred to as the *antiparticle* of the electron and, in general, one assigns the particle/antiparticle labels to other particle pairs in the same way. However, it should be realised that it is purely a convention which of a particular pair one calls the particle and which the antiparticle.

That positrons were not seen in experiments until 1932 is a consequence of a strange and important property of our Universe. Although 'physics' is apparently symmetrical between particles and antiparticles, the Universe is not. In fact, it contains many electrons but almost no positrons. In spite of this it is, to a very good approximation, electrically neutral, i.e. has zero total charge; the negative charge of the electrons is matched by positive charge of other particles (protons). Thus there is a curious asymmetry in the Universe: the negative charge is carried by electrons and the positive charge by protons. Actually, we can only be sure that this is the situation in nearby parts of the Universe—maybe in other parts the negative charge is carried by antiprotons with positrons carrying the positive charge. We return to this asymmetry, which we shall see is crucial to our existence, later.

§9 Rutherford's great experiment

Almost all experiments in elementary particle physics are 'scattering' experiments, in which a beam of particles strikes a target, or in some cases collides with another beam, and the investigator notes what happens. The machine that supplies the initial beam is the 'accelerator' (LEP of §1 is an example); what emerges from the collision is observed by a variety of 'detectors'.

The first great experiment of this nature was performed by Rutherford in 1911. He did not have an accelerator but instead used the naturally occurring 'α particles' that are emitted in the decay of certain radioactive substances (see §14). These particles are in fact what is left when two electrons are removed from helium atoms. The important thing for this experiment is that they have charge $+2e$. Rutherford allowed these particles to fall on a thin metal plate and he measured how many had their paths bent (or were scattered) through a particular angle θ (see figure 9.1). The results he obtained, which looked something like figure 9.2, agreed with what he calculated on

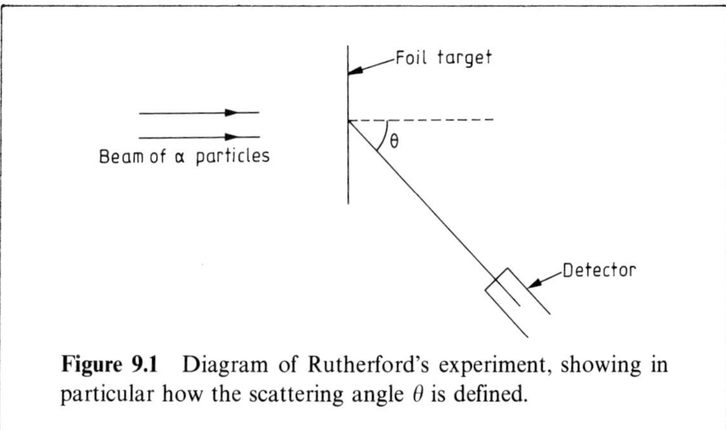

Figure 9.1 Diagram of Rutherford's experiment, showing in particular how the scattering angle θ is defined.

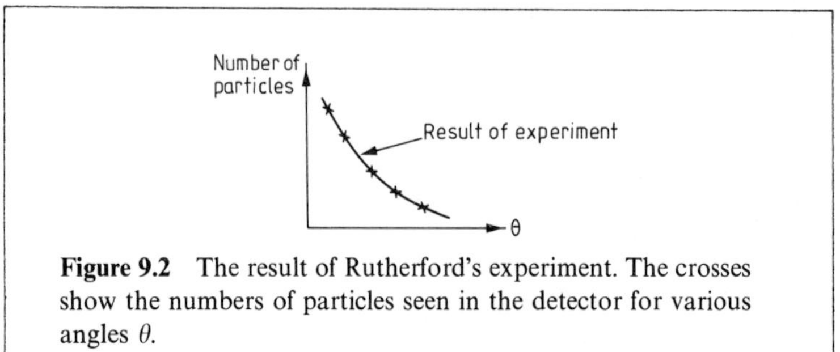

Figure 9.2 The result of Rutherford's experiment. The crosses show the numbers of particles seen in the detector for various angles θ.

the assumption that the target consisted of heavy, positively charged, *point* particles. (This is a reasonably simple calculation given the force law of equation (7.3).) This came as a great surprise! Why?

First, the lack of any effect from the electrons in the metal atoms (recall our currant bun picture) was not unexpected because electrons are so light compared with the α particles ($\sim 1/2000$), and are therefore readily 'swept aside' (just as individual blades of grass cannot significantly affect a golf-ball—sorry, golfers, that excuse won't do!). However, the crucial surprise was that there was no evidence of any effect due to the *size* of the atoms. Recall that the expression given in equation (7.3) referred to *point* charges, and note that it becomes large when the particles are close (r small). For a charge that is spread out over a distance d it is plausible (and can be shown precisely) that the smallest effective separation is d, so that the force is never larger than about Q_1Q_2/d^2. This means that there is much less scattering through large angles than there would be with point charges. Rutherford's results showed clearly that the size of the positive charge distribution in the metal atom was *much* smaller than the size of the atom. In fact we now know that the ratio of the sizes (R_A/R_N) is about 10^5, where R_A, R_N are the radii of the atom and of the charge distribution respectively. Rutherford saw no evidence of nuclear size at all and could only say that this number was large.

Thus we learn about the *atomic nucleus*—a very small particle containing all the positive charge, and almost all the mass, of the atom—and we now have a new picture of an atom as a small, massive nucleus, with the electrons rotating around it and being responsible for the size of the atom (see figure 9.3).

Since the law of force between charges is the same as that of gravity,

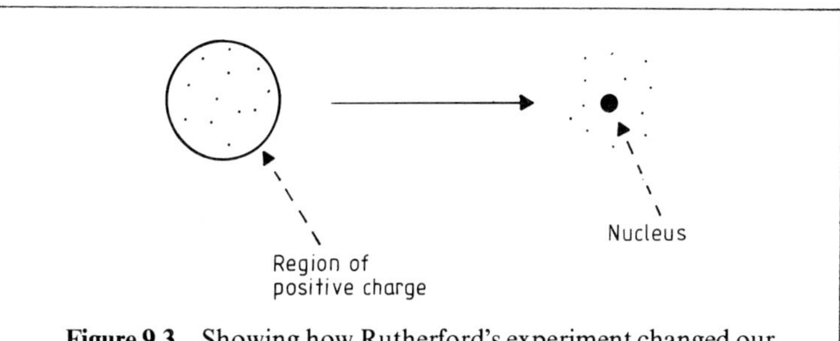

Figure 9.3 Showing how Rutherford's experiment changed our picture of the atom. Electrons are shown as dots. The diagram is not to scale.

as we shall later see, it is clear that in such a system the electrons would rotate in circular, or elliptical, orbits, just like planets around the Sun. Thus each atom would be a miniature solar system—on a scale about 10^{-21} smaller. This is an attractive picture, but it was quickly realised to be seriously deficient.

The first problem with this picture is that all atoms of a given substance appear to be alike. What determines the electron orbits (i.e. their size) and arranges that they are always the same? Even if they are originally 'made' the same, we know that they are frequently colliding with other atoms (unlike, fortunately, the solar system) and would rapidly be disturbed, so that the electrons would plunge into the nucleus. What prevents this? There is actually a much deeper reason why the electron orbits would decay (unless prevented by some other effects), and it is this deeper reason that we shall understand in the next two sections. Then we shall return to the solution of our atomic problem.

§ 10 Electromagnetism

This is another difficult section. The details are not important but the general ideas are so beautiful that no reader will want to omit it! In §7 we met electric charge and electric fields. It turns out (from experiment) that there is another type of field that is caused by, and that affects, electric charges, namely the *magnetic field*, which we shall denote by $\boldsymbol{B}$. The effect in this case only occurs when the charge is moving, i.e. it depends upon the velocity, $\boldsymbol{v}$, of the charged particle.

Let us try to guess what the law giving the effect might be, i.e. what equation determines the force, on a charge, due to the magnetic field? We are looking for something similar to equation (7.5) but we want the velocity to appear. Thus, the right-hand side of the equation will now contain two vectors, $\boldsymbol{v}$ and $\boldsymbol{B}$. We want it to be zero when either of these are zero, so it should contain a product of their magnitudes. We also need a direction, and here the situation is more complicated than in §7 where there was only one vector, $\boldsymbol{E}$, to define a direction. Now we have two: $\boldsymbol{v}$ and $\boldsymbol{B}$. We know, however, that two lines define a plane (unless they happen to be the same line, which case we treat below), and a plane naturally defines a new line, namely, the direction perpendicular to it. Thus an obvious guess for the direction of the force is the line perpendicular to both $\boldsymbol{B}$ and $\boldsymbol{v}$. As noted above, this is not unique if $\boldsymbol{v}$ and $\boldsymbol{B}$ are parallel, so it is natural to suppose that the force is zero in this case. This can be arranged if, instead of taking the products of v and B, we take the product of the magnitude of $\boldsymbol{v}$ and the component of $\boldsymbol{B}$ perpendicular to $\boldsymbol{v}$; we denote this component by $B_\perp$. There is, in fact, a special notation for such a product:

$\boldsymbol{v} \wedge \boldsymbol{B} =$	v	$\times$	$B_\perp$	$\times$	$\hat{\boldsymbol{n}}.$
	Magnitude of $\boldsymbol{v}$		Component of $\boldsymbol{B}$ perpendicular to $\boldsymbol{v}$		Unit vector perpendicular to $\boldsymbol{B}$ and $\boldsymbol{v}$

(10.1)

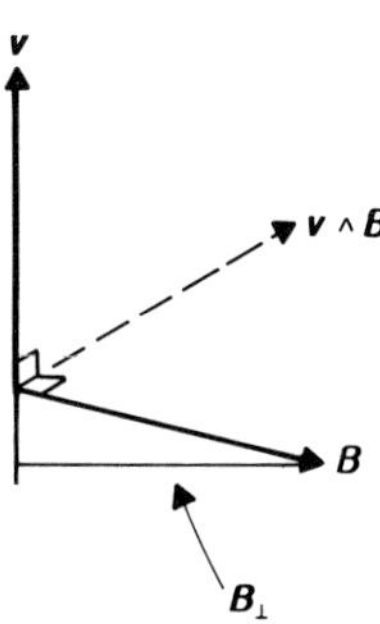

Figure 10.1 Illustrating the vector product. All the lines are in the plane of the paper except $\boldsymbol{v} \wedge \boldsymbol{B}$ which is meant to be perpendicular to it. Readers who are familiar with elementary geometry will easily be able to convince themselves that the vector product is symmetrical in that the magnitude of $\boldsymbol{B} \wedge \boldsymbol{v}$, which is the magnitude of $\boldsymbol{B}$ times the component of $\boldsymbol{v}$ perpendicular to $\boldsymbol{B}$, equals that of $\boldsymbol{v} \wedge \boldsymbol{B}$. From the definition of the direction given in the text it then follows that $\boldsymbol{B} \wedge \boldsymbol{v} = -\boldsymbol{v} \wedge \boldsymbol{B}$.

It is called the *vector product* of $\boldsymbol{v}$ and $\boldsymbol{B}$, and is illustrated in figure 10.1. Then, our 'guess' is

$$\boldsymbol{F} = Q\boldsymbol{v} \wedge \boldsymbol{B} \tag{10.2}$$

a result which is (of course!) correct.

Note, however, that a plane only defines a direction up to a sign, so we need an experiment to determine the sign in equation (10.2). Even then we need some 'convention' to tell us the exact direction of $\hat{\boldsymbol{n}}$. We might, for example, take the first finger of the right hand to point along $\boldsymbol{v}$ and the second finger to point along $\boldsymbol{B}$; then the thumb, placed at a right angle to both, points along $\hat{\boldsymbol{n}}$. That we need a model, e.g. *a right hand*, to specify the direction is an amusing subtlety to which we shall return later.

Next, we need an analogue of equation (7.6), i.e. we need to know how moving charges cause magnetic fields†. This must be such that, when we write down the force between the two moving charges (i.e. by

† Those who know that magnets cause magnetic fields need not close the book. A magnet is, in effect, an organised pattern of moving charges.

eliminating $\boldsymbol{B}$, in the way that we can eliminate $\boldsymbol{E}$ from equations (7.5) and (7.6) in order to obtain equation (7.3)), it is symmetrical between the charges. There are some subtleties here, which we shall ignore, and just write down the result:

$$\boldsymbol{B}=\beta Q\frac{\boldsymbol{v}\wedge\hat{\boldsymbol{r}}}{r^2} \tag{10.3}$$

as the field $\boldsymbol{B}$ at a point P, due to a charge Q at O moving with velocity $\boldsymbol{v}$, where $\boldsymbol{r}$ is OP. This is illustrated in figure 10.2.

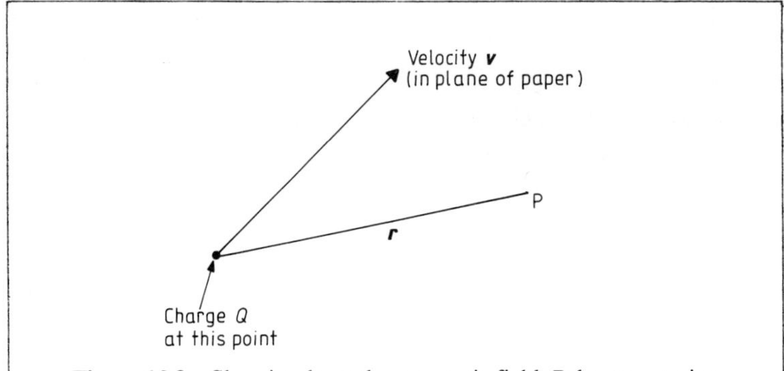

Figure 10.2 Showing how the magnetic field $\boldsymbol{B}$ due to a point charge Q moving with velocity $\boldsymbol{v}$ is constructed. At P the field is perpendicular to the paper and downwards.

The factor β (beta) in equation (10.3) is a constant which has to be determined from experiment. As was the case for the factor α introduced in equation (7.3), the value of β depends upon the unit adopted for the measurement of charge. However, it is important to realise that once α, say, has been fixed by a suitable choice of unit, then there is no further freedom left, and β is a *measurable* quantity. In fact it is easy to see that we can make from α and β a quantity that does not depend on the unit adopted for charge. What sort of a quantity is this? To find out we first use equations (10.2) and (10.3) to deduce that the units of β are† (mass) × (length) × $(\text{charge})^{-2}$. Then, recalling the units for α given in §7, we find that the quantity c defined by

† We shall need to realise that the unit vector $\hat{\boldsymbol{r}}$ has no dimensions. This can be seen easily by writing it as $\hat{\boldsymbol{r}}=r^{-1}\boldsymbol{r}$.

$$c^2 = \alpha/\beta \tag{10.4}$$

does not contain the charge unit. In fact c has units (length) × (time)$^{-1}$ and is therefore a velocity.

Thus, the physical laws of the Universe contain a *fundamental velocity*, c. This is a very remarkable fact and we are led to ask, first, what is it and, second, does anything in the Universe move with such a velocity?

The answer to the first question is

$$c = 2.997\,924\,58 \times 10^8 \text{ m s}^{-1}. \tag{10.5}$$

(We give all the figures to show how incredibly clever experimental physicists are! There is some uncertainty in the last figure quoted.) By comparison with velocities with which we are familiar, this is large; an object moving with such a velocity would go eight times around the Earth in one second!

The answer to the second question is yes. This answer was discovered theoretically by J C Maxwell in 1864, and it represents the greatest ever success of the application of mathematical thought to physical phenomena. Maxwell took the equations we have obtained in §7 and in this section, together with others obtained from experiments done with changing electric and magnetic fields, wrote them in a mathematically elegant and *consistent* form, and then showed that they had solutions that were not envisaged in deriving the equations. These solutions represent *electromagnetic waves* which travel across space with a velocity c. Maxwell correctly suggested that light was such an electromagnetic wave, and that indeed c is the velocity of light. He also predicted other types of electromagnetic waves, and so opened the way to radio waves, x-rays, etc. Clearly, we shall need to know a bit about waves, and to this subject we shall turn in the next section.

Before we leave this section, however, we should pause to note that the theory of electromagnetism represents one of the supreme achievements of the human intellect. The variety of phenomena involving electric and magnetic fields have had a gigantic effect on human activity; it is hard for us now to imagine a world without electricity, electric motors, radio communication, computers, etc. The impact of the theory on scientific thought has also been profound; in one sense it gave rise to quantum mechanics, relativity, gauge theories, indeed all of modern physical theory!

§ 11 Waves

'Vibrations', or 'oscillations', are familiar in everyday life and are very important. In this section we shall find out lots of interesting things about them. Consider, first, a simple pendulum as shown in figure 11.1. This consists of a string attached at one end to a fixed point and at the other to a small weight. The string will hang at rest vertically—a position which is called the *equilibrium* position. If the weight is moved to one side and then released, it will swing back to the initial position. This is due to gravity. However, it will not stop at the initial position but, because of the *inertia* of the weight, will move beyond it, before gravity again pulls it back. The process will continue; in fact, if there were no *damping* due to air resistance, etc, it would continue for ever.

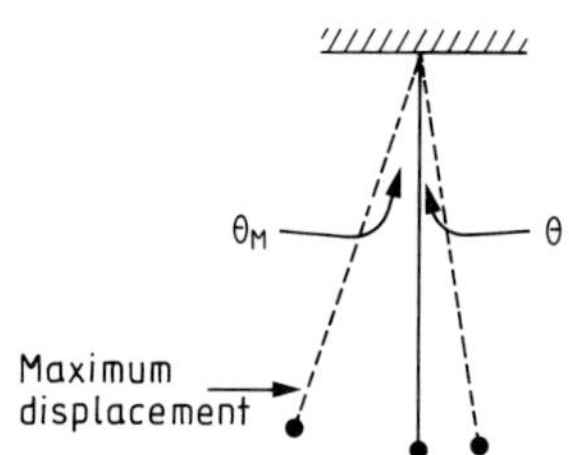

Figure 11.1 A simple pendulum. The angle θ is the angle of deflection from the equilibrium position ($\theta = 0$) and its maximum value is θ_M.

We can define two properties of this oscillation. The *amplitude* is the largest displacement from the initial position. For example, if we use as our measure of displacement the angle between the string and the vertical, the largest value, θ_M, is the amplitude. We see that this value of θ occurs when the weight is instantaneously at rest, i.e. when its velocity is changing direction. Clearly, for a given pendulum, we can have a range of amplitudes, starting from zero. Because of the damping effects mentioned above, the amplitude decays slowly with time.

The other property is the *period*, which is the time taken for the mass to return to its initial position. To obtain a better understanding of this, we draw a graph showing the way the angle θ varies with time. This will look like figure 11.2, where we have assumed that the pendulum is released from rest at an angle θ_M at $t=0$, and have included the effect of a small amount of damping. As indicated in this figure, the period T can be taken between any two adjacent, identical, positions and velocities on the curve.

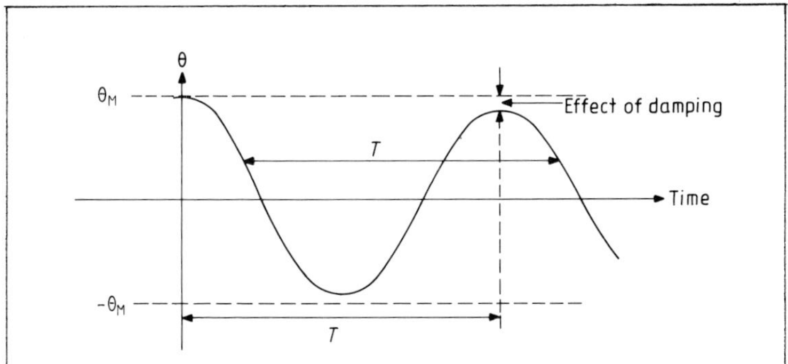

Figure 11.2 A graph of the angle θ, defined in figure 11.1, showing how it varies with time. The period, T, is also shown.

It is more usual to quote, instead of T, its inverse, which is denoted by the Greek letter ν (nu)

$$\nu = T^{-1}. \tag{11.1}$$

This represents the number of times the motion returns to any given state within one second (or within whatever we choose as our unit of time), and is thus called the *frequency*.

A remarkable feature of the pendulum (and of many similar systems) is that the frequency is independent of the amplitude. This can be verified experimentally and, of course, can also be shown theoretically by solving the equations giving the motion of the pendulum. (In fact this result—like many 'nice' results—is only approximately true; it depends on the amplitude being reasonably small, for example.) Thus, the frequency, unlike the amplitude, is an *intrinsic* property of the system—in fact, it depends only on the length of the pendulum.

Now, from a simple oscillating system, where we have a displacement that depends only on time, we want to move to the idea of a *wave*, which is an oscillation where the displacement depends on position as well as on time. For an example, consider a still lake (this is the equilibrium position), and suppose a long thin rod is dropped onto its surface. A wave pattern will move away out from the line of impact and at any given position, the height of the surface water, H, will oscillate like the pendulum (figure 11.2 with θ replaced by H). Moreover, if we consider a line across the surface, starting at a point of impact of the rod and perpendicular to it, and draw a graph of H as a function of distance (x) along this line, we will obtain something like figure 11.3. Notice that the horizontal direction now measures x rather than t as in figure 11.2. In spite of this, the curves have exactly the same form.

We can now define a *wavelength* for our wave; it is the distance we have to move for the function H to return to an initial point. We denote this by the Greek letter λ (lambda); it is illustrated in figure 11.3 and, as with T, it does not matter where, or when, it is measured (provided we ignore the effect of damping).

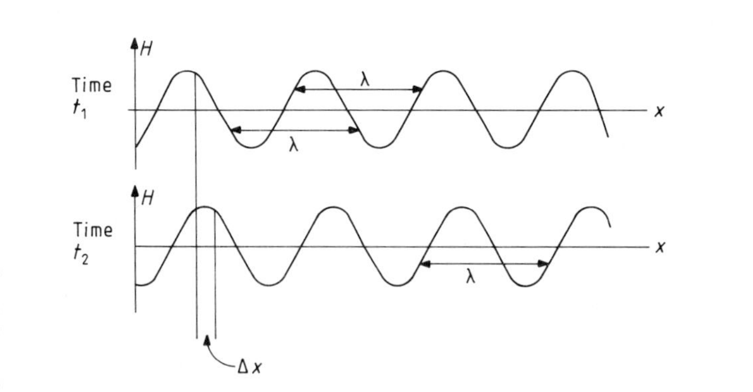

Figure 11.3 The ripple on a water surface, as a function of distance, shown at two times t_1 and t_2.

Figure 11.3 shows the form of H as a function of x at two different times t_1 and t_2, and we see that the wave is moving away from the rod. In order to reach an identical adjacent point on the H curve we must either wait for a time T or move a distance λ, from which (after a few minutes of thought) we realise that the wave is moving across the surface of the pond with a velocity given by the wavelength divided by the period:

$$v = T^{-1}\lambda = \nu\lambda. \tag{11.2}$$

The amplitude, frequency and wavelength can all depend on the nature of the disturbance, but the velocity is independent of these and depends only on the system carrying the wave, i.e. the water.

For water waves the displacement is a function of time and of position on the surface, which is a two-dimensional object (although we only considered a rather special one-dimensional wave in the above). Another very familiar wave motion is *sound*. Here the displacement is *air pressure*, and this is now a function of time and of position in all three-dimensional space. The velocity of a sound wave is again an intrinsic property of the substance carrying it: normally, air at some particular temperature and density. The amplitude of a sound wave determines how loud the sound is. The *quality* of the sound depends on its frequency (most sounds are combinations of many frequencies), which is determined by the frequency of the initial disturbance. A single frequency corresponds to a pure tone which has a particular 'pitch'. The notes of a musical instrument contain, in addition to this single frequency, various higher multiples of it. These distinguish the sounds made by various instruments.

We can now turn to electromagnetic waves. Maxwell discovered that his equations had solutions that represented waves travelling in three-dimensional space. In this case the 'displacement' is the value of $\boldsymbol{E}$ and $\boldsymbol{B}$. These are vectors, so we have a more complicated type of wave than that which transmits sound. In fact, it turns out that, at any given point, $\boldsymbol{E}$ and $\boldsymbol{B}$ are perpendicular to each other and to the direction in which the wave is travelling. The velocity of these waves is, of course, the quantity c. Maxwell correctly deduced that light was such a wave, and it was soon realised that an electric charge that was forced to oscillate with a given frequency would emit electromagnetic waves of that frequency†. Different ranges of frequency tend to be

† This is the 'deeper reason' mentioned at the close of §9 (we have got there at last). Orbiting electrons should emit electromagnetic waves and so lose energy and drop into the nucleus. They don't!

given different names. The BBC Radio 4 'long wave' has $\lambda = 1500$ m, which corresponds to a frequency of 2×10^5 cycles per second or 'hertz' (Hz); a typical 'short wave' is 50 m corresponding to 6×10^6 Hz. We increase frequencies through VHF radio, microwaves, infrared radiation, until we reach visible light, where λ is around 10^{-10} m and ν around 10^{18} Hz. Different colours, of course, are simply different frequencies. Modern accelerators produce electromagnetic radiation with frequencies greater than 10^{25} Hz. All these apparently different phenomena are examples of electromagnetic waves.

We must now learn about a very characteristic property of all waves, namely, *interference.* To understand this, consider again the lake, and suppose that two identical pebbles are dropped into it at different points. From each of these a circular wave will emanate. When the two waves meet we will obtain a very complex pattern. At some positions the waves will arrive so that a crest from one impact will coincide with a crest from the other, and similarly for the troughs; thus we will obtain the same oscillation but with twice the amplitude. At other positions on the surface, a crest from one impact will always coincide with a trough from the other, so that the waves will cancel and the surface will remain undisturbed. This is the phenomenon of interference, which of course can only give a steady pattern if the two interfering waves have identical frequencies.

The most readily observed example of interference of light waves is in the colour patterns of thin films, e.g. oil on any flat surface. Here, light reflecting from the top surface of the oil interferes with that reflecting from the bottom surface. Since the precise positions for cancellation depend on the wavelength, we observe a pattern of changing colour. (The fading of short-wave radio, which is reflected by the ionosphere, is a similar interference phenomenon.)

There is one more topic we must discuss before concluding this section; the idea of *natural* frequencies of wave motions. Recall that, in the case of a pendulum, there is just one frequency of motion, whereas for waves this is not generally the case. However, if the system carrying the waves is in some way constrained, then we again find particular frequencies—usually an infinite number of them. To understand this, consider a new system consisting of a stretched elastic string fixed at its ends. In the equilibrium position, the string will be straight, but if it is disturbed it will oscillate from side to side. However, the end points must remain fixed. This means that they

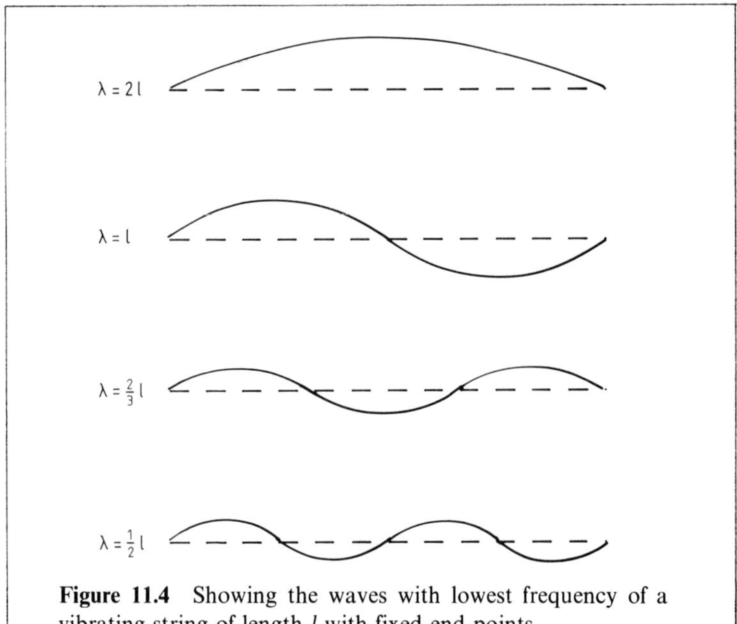

Figure 11.4 Showing the waves with lowest frequency of a vibrating string of length l with fixed end-points.

must always be at the zero-displacement points of the wave. How can this be when the waves are travelling with a fixed velocity? Interference provides the answer; the motion actually consists of two waves travelling in opposite directions which always cancel at the end points. In order that this cancellation can occur at *both* end points the separation between them has to be an exact integral multiple of half the wavelength (half because the wave crosses zero twice in each wavelength—see figure 11.3). Thus our string can only vibrate with a set of fixed wavelengths: $2l$, l, $2l/3$, $l/2$, etc, as shown in figure 11.4. Hence, because of equation (11.2), it can only have a set of discrete frequencies beginning at the lowest $\nu = v(2l)^{-1}$, where v is the velocity of waves on the string. This lowest frequency gives the 'note' of, for example, a violin string. The other frequencies are the overtones and, as noted above, the quality of a note depends on the other frequencies and the amplitudes associated with them. A general motion of a string is a combination of all allowed frequencies.

§ 12 Quantum mechanics

The early part of the century produced two major theoretical developments which form the basis of our understanding of physics: relativity, which we shall meet in §15, and quantum theory, which is the subject of this section.

So far in our discussion we have talked about 'particles', which at any given time would be expected to have a unique position, and 'waves', which have different characteristics, i.e. amplitudes and frequencies. Indeed, a wave of one single frequency has to look like figure 11.3 for all x, so it cannot in any sense be associated with a position. It was, therefore, a great surprise to physicists to realise that this distinction between particles and waves is unreal and applies more to the way we observe an object than to the object itself. (We have some everyday familiarity with this sort of thing: a man can be described in terms of his employment—rank, salary, etc—or, alternatively, in other contexts, he could be thought of purely in terms of the chemicals that make up his anatomy. This analogy, of course, is inadequate, and should not be taken too seriously.)

We describe, first, the experimental evidence for the dual nature. Electrons have been introduced as particles, and indeed they can be observed to make tracks in suitable detectors, and to hit other detectors at particular points. On the other hand, light has been explained as a wave motion according to Maxwell's theory, and it is known to produce interference patterns which are characteristic of waves. So we have, apparently, a clear distinction: an electron is a particle, light is a wave. *The distinction, however, is spurious, since we can do experiments that produce exactly the opposite result.* For example, we can scatter electrons off suitably ordered arrangements of atoms† and obtain interference patterns, thereby showing that

† A crystal provides such an arrangement. Each atom now represents a 'source', so that this is similar to the two-pebble experiment of §11, except that there are more than two pebbles and the pattern is more complicated.

electrons are waves. Similarly, if we allow electromagnetic radiation to fall on a metal foil, the energy is not distributed uniformly across the surface of the foil; instead, electrons are knocked out of individual atoms exactly as would be expected if the radiation was a beam of particles. Also, electromagnetic radiation is not emitted and absorbed continuously but in discrete 'lumps', i.e. the radiation behaves as particles.

This particle–wave duality is described by two relations. The first can be obtained by using radiation of a fixed frequency, ν, in the second experiment of the previous paragraph. Then we find that the energy of the 'particle' that hits the atom is a constant multiple of its frequency, thus

$$E = h\nu \tag{12.1}$$

where h, Planck's constant, is a fundamental constant of nature. From this equation we see that its units are energy multiplied by time, i.e. $(\text{mass}) \times (\text{length})^2 \times (\text{time})^{-1}$.

For the other relation we need first to define the momentum of a particle. This is a vector, $\boldsymbol{p}$, in the direction of the velocity and with magnitude equal to that of the velocity multiplied by the mass of the particle. Thus:

$$\boldsymbol{p} = m\boldsymbol{v} \tag{12.2}$$

for a particle of mass m moving with velocity $\boldsymbol{v}$. (Actually this relation is only approximate—the exact form is given in §16—but the error is small provided v is very much less than c.) Then the magnitude of $\boldsymbol{p}$ is proportional to the inverse of λ:

$$p = h/\lambda \tag{12.3}$$

where the same constant h appears. Such a relation can be verified by analysis of the electron interference experiment described above.

In order to illustrate some features of this dual nature we imagine the 'two-slit' experiment shown in figure 12.1. Considered as *waves*, the objects emitted by the source split into two parts, one passing through slit A and the other through slit B. Because the path lengths, e.g. AP and BP, are different (in fact, approximately, AP − BP = 2(AB)(OP)), there will be an interference pattern. This means that some points P will be 'dark', i.e. will not receive any particles. However, a particular particle that does arrive at the screen will, considered as a particle, have passed through one or the other slit—it

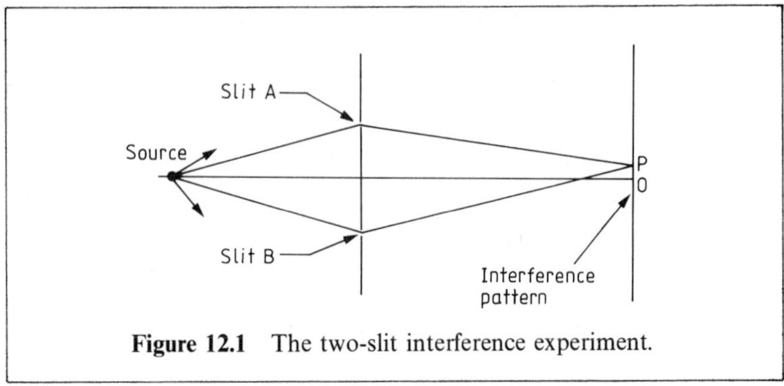

Figure 12.1 The two-slit interference experiment.

cannot have passed through both. The other slit, however, is not irrelevant because it determines the possible dark positions P. Long discussions on whether there is, or is not, an essential paradox here have been, and still are, a part of the history of quantum mechanics. At this stage we need only note that the theory always allows us to make predictions which agree with experiment. (The 'problems' come when we try to think too hard about what we are doing. We postpone such thinking to our final section.)

We now describe several consequences of quantum theory which we shall use later.

(*a*) As we noted above, a wave with a unique wavelength has *no* position. How can we make a wave with a position? We must combine many waves with different wavelengths, and try to arrange that they interfere in a suitable way; that is, at, or near, a particular point they must add so that we obtain a large amplitude, whereas in all other regions they cancel each other so that we have a very small amplitude. There are some nice theorems that tell us that we can indeed make such a 'wave packet', provided we use a continuous range of wavelengths. This is illustrated in figure 12.2. In picture (*a*) we see a wave with 'wavelength' λ_0, but with a varying amplitude, rising from zero and then falling off. The size of the packet is written as Δx. This can be constructed by adding waves of unique wavelengths (like that of figure 11.3), with a smoothly varying amplitude, as shown in figure 12.2(*b*). The amplitude is largest near the wavelength λ_0. We have written $\Delta\lambda$ as the 'spread' of wavelengths required. As we might expect, the smaller we make Δx the larger we have to make $\Delta\lambda$. There

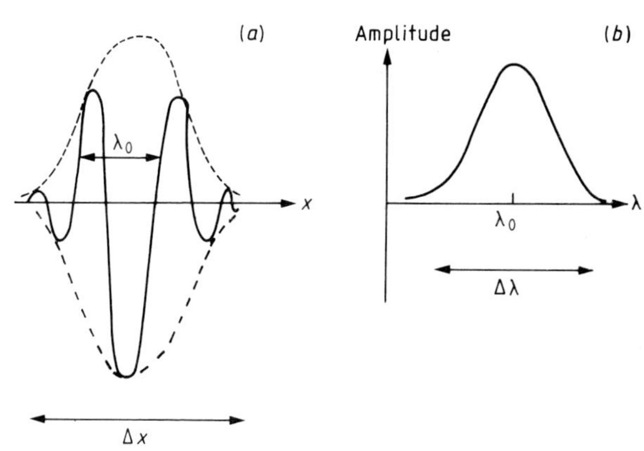

Figure 12.2 Showing how a wave packet of width Δx and 'wavelength' λ_0 can be made of waves, extending over all space (all x), with a range of wavelengths in a neighbourhood $\Delta\lambda$ about λ_0.

are theorems of mathematics which put this statement in a precise form. It is, in fact, more convenient to consider the spread in values of p (related to λ through equation (12.3)), and these theorems then tell us that the product of the spread of x with the spread of p is always at least h:

$$(\Delta x)(\Delta p) \geqslant h. \tag{12.4}$$

The symbol $\geqslant$ means that the product is greater than, or equal to, h; the equality only occurs when the packet has a particular shape. This equation is *Heisenberg's uncertainty principle.* Since Δx is a measure of the uncertainty in the position of our object and Δp is a measure of the uncertainty of its momentum, the equation tells us that the more accurately we determine the position, the bigger will be the error in the momentum, and vice versa. This is a 'can't win' theorem; what you 'gain on the roundabout' you 'lose on the swings'; the more you know about the position, the less you know about the momentum.

At this stage readers may well be protesting that this must be nonsense. It is utterly contrary to our experience. We *know* that we can measure the position and velocity (and hence momentum) of an object to any desired degree of accuracy. The reason why this, apparently quite reasonable, protest need not disturb us lies in the extremely small value of h when measured in units appropriate to everyday experience. In fact

$$h=6.6262\times10^{-34}\ \mathrm{kg\,m^2\,s^{-1}} \tag{12.5}$$

so, for example, if we measure the position of a mass of one gram to an accuracy of 10^{-5} m, then the velocity uncertainty is about 10^{-25} m s^{-1}, corresponding to the particle having moved 10^{-8} m in a time as long as the age of the Universe! *Quantum effects in everyday experience are negligible—they are important in atoms because of the very different magnitudes involved.*

(*b*) Since all particles are associated with waves, the phenomenon of discrete frequencies discussed at the end of the previous section can arise. This discreteness in fact was one of the main characteristics of early work using quantum theory, and gave rise to the name. We shall see how this occurs in atoms in the next section. Another important example is the *spin* of elementary particles. The *angular momentum*† associated with spin has to have the value $(n/2)(h/2\pi)$, where n is zero or a positive integer. A given particle always has a fixed value of n. Thus an electron always has $n=1$ (we tend to drop the $(h/2\pi)$ and say an electron has spin $\frac{1}{2}$). The particles associated with electromagnetic radiation are called 'photons' and these have $n=2$ (hence a photon has spin 1).

Angular momentum and spin are *vector* quantities, because, for example, the axis about which the object spins is a line in space. (Actually to obtain the *sign* of the spin along this line requires some model, e.g. a right-handed screw, so something like the discussion below equation (10.2) is involved again here.) We can thus discuss the components of the spin in any given direction. Here comes a surprise: they are also 'quantised' and can only take the values $(\pm n/2)(h/2\pi)$, where again n is zero or an integer. In fact, for a spin $\frac{1}{2}$ object like an electron the components of the spin in any chosen direction are one of the two values $\pm\frac{1}{2}$. For a spin 1 object on the other hand the values ±1 or zero are generally allowed, whereas spin 2 permits ±2, ±1 or zero, as illustrated in figure 12.3.

† This is the rotational analogue of momentum defined previously. Its exact definition need not concern us. Note that the number 2π is the ratio of the circumference of a circle to its radius; it is just above 6.

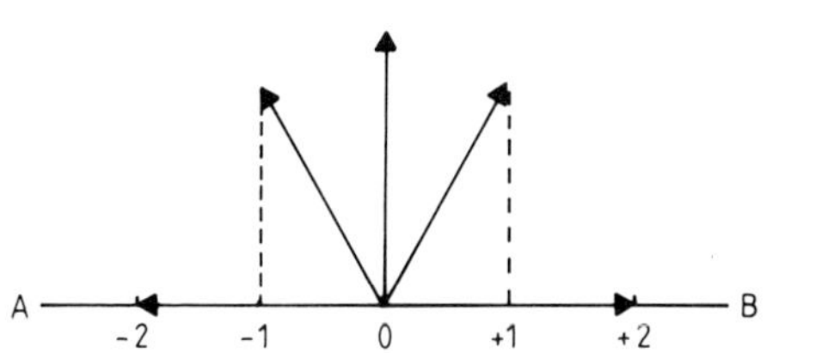

Figure 12.3 Showing possible values for the spin components along AB of a spin 2 particle. In a non-quantum situation the component could be any number between -2 and $+2$.

(*c*) Whereas we are tempted to regard two *particles* of the same type as being distinguishable objects—we tend to think that we might call one Charlie and the other Peter for example—if we think of them as *waves* the difference between having one or two, or more, is simply a matter of the amplitude; no sense of distinguishing separate entities arises. What this means is that the way we define a system must be such that nothing changes if we interchange two identical particles; in other words, the system must be symmetrical between them. Now, in fact, all measurable quantities turn out to depend on the square of the wave amplitude, so sign changes are allowed. For rather complicated reasons it turns out that, whereas for spin 0, 1, etc, objects, the wave remains the same when any two are interchanged, for spin $\frac{1}{2}$, $\frac{3}{2}$, etc, it *must* change sign. This property has a remarkable consequence: we cannot have two electrons (for example) in exactly the same state. (Proof: If we try, then clearly nothing would happen when we interchanged them—so we could not get the required sign change.) This consequence is called the *Pauli exclusion principle.* Note that the states have to be *exactly* the same. Thus, for example, the number of electrons we could put at a particular *position* would be two—we would point their spins in opposite directions so that one had spin projection $+\frac{1}{2}$ and the other $-\frac{1}{2}$, with respect to some arbitrarily chosen axis.

(*d*) Although not actually relevant to our story (except possibly the last section) we must mention here a remarkable feature of quantum mechanics: it destroys the narrow concept of a mechanistically determined Universe, with its future forever calculable from knowledge of its present by the laws of mechanics. The uncertainty principle ensures that even the initial conditions required to specify the future motion of a single particle cannot be precisely stated. Although the wave function describing an isolated system does change in a calculable way, such a wave function only gives probabilities for values of measurable quantities. The significance of this breakdown of determinism, for our understanding of the world and our place in it, has been, and will remain, a fascinating subject of discussion. We shall briefly return to it in the final section.

This has been a long section—quantum theory is a big subject. I hope that some of the ideas will become clearer as we use them in what follows. We are now, at last, in a position to understand atoms.

§ 13 The structure of atoms

First, let us try to solve the problem posed at the end of §9, namely, why the electrons which orbit the nucleus do not fall down into it (through collisions or by emitting electromagnetic radiation). The point is that a closed orbit is rather like the string discussed in §11, in that it can only have waves of particular frequencies. It is not here a question of the wave being zero at the end points; rather, when we have moved all the way round the circumference we must arrive back at the same place, i.e. the circumference (we take a circular orbit for simplicity) must be an integral multiple of the wavelength (see figure 13.1). We shall now do a simple calculation to show that only certain orbits are allowed and that these have particular energies $E_1 < E_2 < E_3$, etc. Readers who do not wish to follow the details of the calculation should jump to equation (13.7) where they will find the result.

The equation which balances the electric force between the nucleus of charge Ze and the electron ($(Ze/R^2)e$ where R is the radius of the orbit) with the centrifugal force due to the orbital velocity is

$$\frac{Ze^2}{R^2} = \frac{m_e v^2}{R} \tag{13.1}$$

where m_e is the electron mass. So the square of the momentum (recall equation (12.2)) is given by

$$p^2 = (mv)^2 = \frac{m_e Ze^2}{R}. \tag{13.2}$$

Using equation (12.3) we obtain for the wavelength λ

$$\lambda^2 = \frac{h^2 R}{m_e Ze^2}. \tag{13.3}$$

The condition for a satisfactory orbit is, as noted above, that

$$n\lambda = 2\pi R \tag{13.4}$$

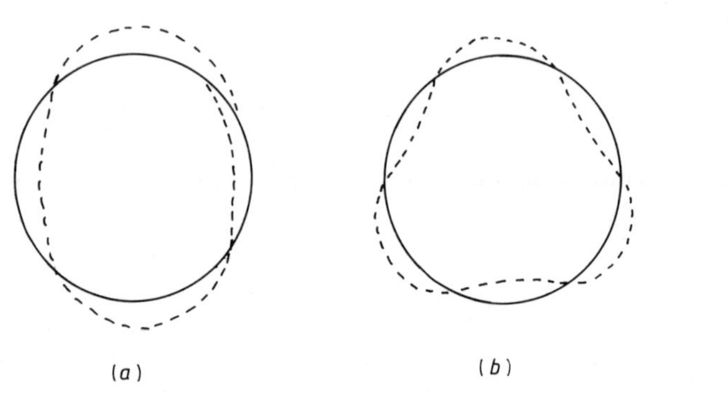

Figure 13.1 Showing a wave which does not fit into a given circle (*a*) and one which does (*b*). In the latter case the wavelength is one third of the circumference.

(this is the circumference of a circle of radius R) when n is again an integer, or

$$\lambda^2 = \frac{(2\pi R)^2}{n^2}. \tag{13.5}$$

On comparing the two expressions for λ^2 given by equations (13.3) and (13.5) we obtain

$$R = \frac{h^2 n^2}{(2\pi)^2 m_e Z e^2}. \tag{13.6}$$

The energy of the electron in its orbit is the sum of its kinetic energy ($\frac{1}{2}m_e v^2$) and its potential energy ($-Ze^2/R$) so equation (13.1) gives $E = -(e^2 Z/2R)$, and using equation (13.6) we finally obtain for the energy of the nth state

$$E_n = -\left(\frac{Z^2 e^4 m_e (2\pi)^2}{2h^2}\right)\frac{1}{n^2}. \tag{13.7}$$

This simple formula gives, with $Z=1$, exactly the observed energy levels of hydrogen. The lowest state, with $n=1$, is the normal state of a hydrogen atom and its size is then given by equation (13.6) with $Z=1$,

$n=1$ as

$$R=6\times10^{-11}\ \text{m}. \tag{13.8}$$

The states with n greater than one are called 'excited' states. They can be observed, for example, when the gas is heated. This causes the atoms to move with higher velocities, so that an atom can be put into an excited state by collision with another atom. Such an excited atom will then 'decay' into a lower energy state, and eventually into the lowest state ($n=1$), by emission of electromagnetic radiation (light). Because energy is conserved, a given decay emits radiation with an energy which is the difference between the values given in equation (13.7) for different values of n. Thus, recalling the relation between the frequency and energy, equation (12.1), we find for the frequencies emitted by the hydrogen atom

$$\nu=\left(\frac{e^4 m_e (2\pi)^2}{2h^3}\right)\left(\frac{1}{n^2}-\frac{1}{m^2}\right) \tag{13.9}$$

for various integral values of n and m. It was the agreement of this quantum mechanical expression with the (very accurate) experimental observations that helped to persuade even sceptical physicists of the truth of quantum mechanics. Note that the atom in its lowest state, $n=1$, cannot radiate because the electron cannot lose energy—it already has the lowest *possible* energy. This, of course, is the solution to our problem at the end of §9.

In other atoms the situation is much more complicated. One effect of the higher charge on the nucleus can easily be accommodated by inserting the appropriate value of Z in the above equations, but we must also include the electric interactions of the electrons with each other and take account of the exclusion principle (§12). Even for hydrogen our small calculation is oversimplified—it ignores, for example, non-circular orbits. However—and this is the great triumph of quantum mechanics—*there is a simple equation† resulting from the theory which uniquely and correctly predicts the properties of atoms.* The way in which the electron states arrange themselves to give the rich variety of observed atomic physics is itself a fascinating story. In many cases the calculations are too complicated to do except in a very approximate way—even with large computers—but nobody doubts that the right results would be obtained—the 'errors' are due to

† First given by E Schrödinger and hence called the Schrödinger equation.

inadequate calculational techniques, not to the theory.

This applies not only to the state of individual atoms, but also to the forces between them, to molecules, and hence to the whole of chemistry, to the properties of matter and to all the forces with which we are familiar in the macroscopic world. *In principle*, these things are now reduced to the solution of a single known equation. Of course it requires great ingenuity to 'solve' this equation in an approximation suitable for the problem being considered. From the point of view of fundamental physics, however, these subjects are solved—they have nothing more to tell us.

Before we leave them, however, we take a look at a relatively simple problem, namely, the force between two hydrogen atoms at a certain distance, r, apart (take r very much larger than R, the radius given by equation (13.9)). Strictly speaking, this problem is not really *two hydrogen atoms*, since the hydrogen atom is a well defined problem with two charged particles. Our problem now has four charged particles, each interacting with the others, so we have a completely new problem to solve. (Note that in general this is why molecules need not behave like their constituent atoms—a fact noted in §6.) However, if r is much bigger than R we might expect that, in some approximation, we can begin by solving each atom separately and then try to modify the situation because of the presence of the other. At first sight, after we have solved the two hydrogen atom problems, we have *no* residual force between them because each atom is electrically neutral, i.e. its total charge is zero. However, this is not quite correct because the atoms are not 'rigid'. Consider for example the two nuclei, which are much heavier than the electrons, as *fixed*. Then the effect of the repulsion between the electrons will be to cause their orbits to be slightly non-circular so that *on average* they are pushed further away from each other than if the orbits were the original circles. This slightly reduces the repulsion between the electrons, and so gives rise to a net overall *attractive* force between the two atoms. A precise calculation of this effect can be made and it leads to the so-called van der Waals force†, which in fact falls off like r^{-6} (compare r^{-2} for the electric force).

This brings us to the end of the first chapter of our story. We have moved a long way so, for readers who might be a little puzzled about 'where we are', we shall add a summary.

† Generally between *molecules*, not atoms, but the idea is the same.

Summary of Chapter One

Matter is made of atoms of which there are 92 different kinds (§§4, 6) and, in principle, we can calculate the physical and chemical properties of all matter from knowledge of the properties of atoms. Each atom consists of a small nucleus surrounded by electrons (§§8, 9). The type of atom is determined by the value of the positive charge on the nucleus, which is equal to the number of electrons in the atom. All the properties of an atom can be deduced from the rules of quantum mechanics (§12), together with the very simple electric force law which was obtained from laboratory experiments. There are only two input parameters needed, h and e. Where calculations are possible, the theory gives essentially perfect agreement with experiment. The *only* other input is the existence of 92 types of nuclei with particular masses and charges 1, 2, 3 etc. It is these 92 nuclei that must be studied further in the next chapters. Can we be as successful in understanding these, as we have been with atoms?

Chapter Two

in which we pose the question of the origin of nuclei,
we learn new things about space and time,
we meet quantum field theory and its many problems,
we see why gauge theories are so attractive,
we introduce quarks,
and learn how they solve the problem of nuclear forces,
so that we almost reach the end of physics.

§14 The atomic nucleus

At the centre of every atom is its nucleus. It is the electric charge on this nucleus that determines the nature of the atom; hydrogen has a charge 1 nucleus (called a 'proton'), helium has a charge 2 nucleus (called, as we have seen, an α particle), lithium has a charge 3 nucleus, etc, up to uranium with a nucleus of charge 92. Here, to avoid having to include the factor e, we are measuring charges in units such that the electron has charge -1. It is possible to make nuclei with slightly higher charges but these do not occur naturally on earth because they are very unstable. All nuclei have radii which are small compared to atomic radii (by a factor of about 10^{-5}) and which increase with charge. On the other hand their masses, which also increase with charge, are almost as large as the corresponding atomic mass.

At this stage we could make the assumption that these nuclei are 'elementary', or 'God-given', objects. This would essentially be the end of our story. It would not, however, be a satisfactory ending. There are really too many of them and their different radii and masses would all be input constants of our theory. They also have other properties which would not be calculable. In particular, they have complicated interactions with each other, and some of them spontaneously 'decay', i.e. break up into other nuclei.

Fortunately we have good reason for believing that we have not reached the end of our story, namely, that there is an obvious way of continuing it. The observed charges, together with the fact that masses and radii increase with charge, suggest that all nuclei are made of the simplest, i.e. the proton, with the number of protons being equal to the charge. Thus the α particle would be two protons, the lithium nucleus would be three, etc. This does not quite work however; it would suggest masses of twice the proton mass for helium, 3 times for lithium, and so on, whereas actually the factors are, approximately, 4, 7, etc. The reason for this is that there is another particle, the *neutron*, which is very like the proton and, in particular, has almost the same mass, but which has zero charge. The neutron was first observed as a free particle in 1932. Nuclei are made of protons and neutrons, the charge being determined by the number of protons and the mass by the number of protons plus the number of neutrons. The contents of some typical nuclei are shown in table 14.1. In this table the charges are given in multiples of minus the electron charge and the masses in multiples of the proton mass. Actually the quoted masses are only approximate; first, because the neutron mass is not exactly equal to the proton mass, and also for a more important reason which we shall discuss in §16.

Table 14.1 Some typical nuclei and their proton and neutron content.

Element	Name of nucleus	Charge	Mass	Content
Hydrogen	proton	1	1	1 proton
Helium	α particle	2	4	2p+2 neutrons
Lithium		3	7	3p+4n
⋮				
Oxygen		8	16	8p+8n
⋮				
Iron		26	56	26p+30n
⋮				
Lead		82	207	82p+125n
⋮				
Uranium		92	238	92p+146n
Deuterium	deuteron	1	2	1p+1n
Tritium		1	3	1p+2n
Helium 3		2	3	2p+1n
etc				

Table 14.1 shows, in addition to the most common forms of nuclei, others with the same charge but different numbers of neutrons. They are called isotopes and often occur naturally, though they are rarer than the main sequence. Note that, from the point of view of atomic physics and of chemistry, where only the charge is relevant, the deuterium and tritium nuclei, for example, are not different from the proton. Any of them could be the nucleus of a hydrogen atom. In fact, hydrogen as found on Earth, e.g. in sea water, contains a small proportion of deuterium and an even smaller proportion of tritium.

Now—what progress have we made? We have a model of nuclei in which they are composites of two types of particle—a proton and a neutron. Apart from seeming to permit the alchemists' dream (gold and iron are actually made of the same things), this is clearly real progress. We appear now to have a very simple world made of just electrons, protons and neutrons. Of course we should remind ourselves here that, as discussed in §8 for the electron, there are antiparticles of both the proton and the neutron, namely the *antiproton* and *antineutron*, with charges -1 and 0 respectively. Although the neutron and antineutron have the same charge they are genuinely *different* particles. Just as matter is made of electrons, protons and neutrons, it is possible to have 'antimatter', made of positrons, antiprotons and antineutrons. This does not occur naturally on Earth; whether it occurs anywhere else in the Universe is a problem to which we return in the last chapter.

In order to confirm our model of nuclei we have to see whether they really do look like composites of protons and neutrons. We must also think about the origin and nature of the forces that hold these constituents together; forces that must be strong enough to overcome the electric repulsion between the protons, and which have to explain why only particular combinations of protons and neutrons occur, and why the nuclei interact with each other in the way that they do.

The first of these questions is that studied by experimental nuclear physics. An immediate success is that our picture offers some understanding of the *decay* of nuclei. As noted before, some nuclei, especially the rarer isotopes, tend to split spontaneously into each other. One form of such decay is the emission of α particles, which we can write as

$$[Z, N] \rightarrow [Z-2, N-2] + \alpha \text{ particle} \tag{14.1}$$

where $[Z, N]$ means the nucleus with Z protons and N neutrons. Using the fact that, in this notation, the α particle is $[2, 2]$, we see that the reaction does not involve one particle changing into another; rather it requires a simple rearrangement of protons and neutrons: two of each from the original nucleus combine to make an α particle which then moves away. A similar rearrangement occurs in *fission*, where a nucleus splits into two, roughly equal, nuclei with the emission of some neutrons:

$$[Z, N] \rightarrow [Z', N'] + [Z'', N''] + k \text{ free neutrons.} \tag{14.2}$$

Here the integers satisfy

$$Z = Z' + Z'' \tag{14.3}$$

$$N = N' + N'' + k. \tag{14.4}$$

A typical example is the fission of uranium ($Z=92$) into barium ($Z=56$) and krypton ($Z=36$). Reactions of this type are involved in the chain reaction which takes place in nuclear reactors; the free neutrons which are released collide with other nuclei and trigger further reactions, releasing more neutrons, and so on. In order for such a reactor to be self-sustaining it is necessary that the material in which the reaction occurs is above a certain critical size, otherwise too many neutrons escape.

The convenient way to quantify the likelihood of the various decays taking place is through the 'half-life'. According to quantum mechanics it is not possible to say when a given nucleus will decay; we can only speak about the probability of its decaying in a given interval of time. So we suppose that we have a large number of identical nuclei and define the half-life as the time taken for half of them to have decayed. Thus, within a particle's half-life there is a 50% probability of its having decayed. Experimentally measured lifetimes range from tiny fractions of a second up to millions of years.

Further progress on the experimental questions required the use of particle accelerators, which were first made for this work. Why do we need to have high velocity particles to do experiments? There are several reasons but here we mention an important one. In order to 'see' the structure of an object on a scale of distance about d, say, we need a 'probe' (e.g. another particle) with a wavelength about equal to d or smaller. We express this as

$$\lambda \lesssim d. \tag{14.5}$$

Hopefully, this is 'sort-of-obvious'; it can, of course, be shown precisely. If we recall $\lambda = h/p$, we see that the inequality in (14.5) requires that p be equal to, or larger than, h divided by d:

$$p \geqslant h/d. \tag{14.6}$$

Thus, the *smaller* the structure, the higher the required momentum and hence the *bigger* the machine!

With these machines it is possible to observe what happens when high energy protons are scattered by other protons. In the same way they can be scattered by heavier nuclei. We expect the effects will be very similar since the heavy nuclei are just collections of protons. (One must also include the effect of the neutrons, but these are, from this point of view, very like protons.) These experiments amply confirm the picture of a nucleus as a composite of protons and neutrons.

Before we study the theoretical problems associated with the forces that hold protons and neutrons together, we must digress into another exciting branch of theoretical physics—the subject of our next section.

§ 15 Special relativity

Here, we shall introduce the second major theoretical development of this century. To begin, let us suppose that I wished to tell my readers the position of Birmingham. If I stated that it was 10 kilometres east they would consider that I was being unhelpful, not only because they probably do not care about the location of Birmingham, but because the statement carries no information unless I specify a starting point. From *where* is it 10 kilometres east? A position has to be quoted relative to some particular reference point, and there is no obvious origin—*space* does not have a unique starting point.

It is slightly less obvious, but equally true, that the same considerations hold for velocity. It is a meaningless statement to say that an object is moving at 40 kilometres per hour due north, unless I specify the zero-velocity reference point. Of course, if I make this statement about a train, a reader would naturally *assume* that I meant relative to the surface of the Earth at the position of the train. To an 'observer' on the Sun, however, the velocity would be totally different since it would include the effects of the Earth's rotation and its movement about the Sun. Similarly, if I described a passenger on the train as standing still, I would mean that he had zero velocity *relative to the train*—and hence a velocity of 40 km h^{-1} relative to the Earth, etc. As with position, there is no natural zero of velocity and only *relative* velocities are meaningful.

Now let us think about waves. In §11 we discussed, for example, water waves and waves on a string, and we recall that these had a velocity, which was a property of the medium. Naturally this velocity would be measured *relative to the medium* that carries the wave (water, string, etc). If we put the string on a train, then an observer by the track would see waves moving with a different velocity (larger or smaller according to the direction of the wave relative to the train's velocity).

This is all fairly obvious and trivial. However, we should now realise a big problem. What about electromagnetic waves? We glibly said these travel with velocity c—but, *to what is this velocity relative*? There is now no 'medium' which is doing the vibrating, so which observer measures the velocity c: one on the Earth's surface, on a train, on the Sun, or where? We can obtain a good illustration of the problem if we think of light from the Sun falling on the rotating Earth, as shown in figure 15.1. Because of the effect of this rotation, the observer at A should measure a larger velocity for the Sun's light than the observer at B. However, our discussion of electromagnetism did not mention Suns and rotating planets, etc; Maxwell did a calculation which gave the result that *both* observers would measure the *same* velocity (c). It is as if a man standing still, and a man riding a bicycle along the side of the track, both find the train passing them at the same velocity!

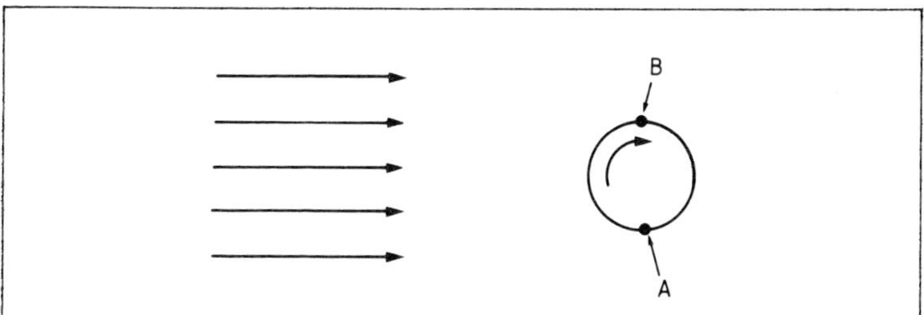

Figure 15.1 The points A and B on the rotating Earth clearly have different velocities relative to some fixed medium. Hence, if light is a vibration of this medium, the difference should be seen in light coming from the Sun.

This, my readers are saying, is not believable. They are then part of a good tradition. Most theoretical physicists, around the turn of the century, would have agreed with them. Electromagnetism must be, in some way, inadequate. A medium, called the 'ether', was invented to carry the electromagnetic vibrations. The value c would then be a velocity relative to the ether, so in figure 15.1 the two observers would certainly see evidence of their velocity relative to the fixed ether. It should be noted that this ether had to be given remarkable properties—for example, since light travels as easily through a 'vacuum' as through air, the ether had to fill any so-called vacuum.

The motivation for believing in such an ether was, at the time, even stronger than that which came from the considerations we have given: it was the great era of 'mechanics'; abstract waves were not fashionable; everything had to be described in terms of mechanical objects—hence the need for a mechanical ether. I have on my desk at this moment notes of lectures given at Baltimore by Sir William Thomson in 1884, in which he is very scornful of attempts to explain electromagnetism without *mechanical* ideas: 'a real matter between us and the remotest stars I believe there is, and that light consists of real motions of that matter.'

The experiment to measure the different velocities at A and B in figure 15.1, and hence to measure the velocity of the Earth relative to the ether, was done by Michelson and Morley in 1887. Actually, it was done in a slightly different way, but this need not concern us—we merely remark that the expected effect was small (owing to the large value of c) so the experiment was not simple. *The ether drift was not found—Maxwell's prediction was seen to be correct*!

At first sight, as with quantum theory (recall the discussion at the end of §12(a)), this appears to violate our experience, in this case regarding the addition of relative velocities—we expect observers moving relative to each other to see different velocities. The explanation is again that the *new* effects are small in the range of velocities with which we are familiar, namely, velocities very much smaller than c.

To appreciate this, let us think how we might *measure* a velocity. For example, suppose we wish to time a hundred metres race. One way of doing this would be to have a clock at the finish which could be started when the race is observed to start and stopped as the winner breaks the tape. This, of course, introduces a small error because of the time taken by the light, which carries the message that the race has started, to reach the finish. (In practice this signal would now be carried electrically, but the velocity is still essentially c.) The error is, however, negligible, because this time is so small compared with the time taken by the runners. Of course, if we were concerned with measuring high velocities (or working to greater accuracy) then we would need to correct for it. However we could not do this until we had measured the velocity of light—and how would we set our clock for this measurement? We have a circular problem: we cannot measure the velocity of light until we know the velocity of light, and until we have measured the velocity of light we cannot measure any velocity!

It was Einstein, in his theory of special relativity published in 1905, who finally sorted all this out. He realised that Maxwell's equations must be correct for *all* observers, who will thus all find the same value, c, for the velocity of light. A consequence is that the usual rule for comparing velocities measured by observers moving relative to each other cannot be correct; it must be an approximation suitable only for velocities which are small compared with c. It is, in fact, easy to derive the correct formula from the fact that the velocity of light is constant. We give the appropriate result in the next paragraph for the benefit of readers who wish to have a few more details.

Suppose one observer measures the velocity of a particular object to be u in some given direction. Then another observer, moving relative to the first with a velocity v, *in the same direction*, will see the object moving with a velocity w given by

$$w = \frac{u - v}{(1 - uv/c^2)}. \tag{15.1}$$

We note two important properties of this formula. For small velocities the quantity uv/c^2 is much less than 1 so it can be ignored, in which case the usual ('obvious') result, $w = u - v$, is obtained. On the other hand, if the observed object is a photon, then $u = c$ and the formula gives $w = c$ also, regardless of the value of v. Hence, as required, both observers measure the same value for the velocity of light.

Another consequence of special relativity is that distances and time intervals are not necessarily the same when measured by different observers. To understand what is involved here, consider measuring the length of a rod. This is straightforward for an observer at rest relative to the rod. However, for a second observer, moving with respect to the first, the rod will not be at rest, so in order to measure the length it is necessary to record the positions of the end points *simultaneously*, i.e. at identical times. This brings us back to the same question which was involved in measuring velocities; we need a method of synchronising clocks at different space points. It is when we do this with light signals, moving always with velocity c, that we discover that the measured length depends on the relative velocity of the observer.

A beautiful feature of special relativity is that this observer-dependence of distances and time intervals does not apply to a particular combination of them, given explicitly below, for which the changes exactly cancel and which is therefore an *invariant*. The easiest way to understand this, and the best way to express the ideas of special relativity, is to extend the discussion of space given in §5 by taking time as an extra 'dimension'. Then changing from one observer to another, with a non-zero relative velocity, is like making a rotation of the axes in this 4-dimensional space.

An illustration of this can be seen in figure 15.2 where we show time and one direction of space. We suppose that the first observer uses the axes labelled x and t, and the second observer uses the rotated axes represented by dashed lines and labelled x' and t'. We take any point, P, fixed relative to the first observer. Such a point will have a constant value of x, labelled x_P in the diagram. Its path will therefore be as shown by the heavy line. Now consider the motion of the point P as seen by the second observer. At $t'=0$, it is at the position labelled $x'=x'_P$, whereas, at $t'=t'_1$, it is at $x'=0$. Thus P has a (negative)

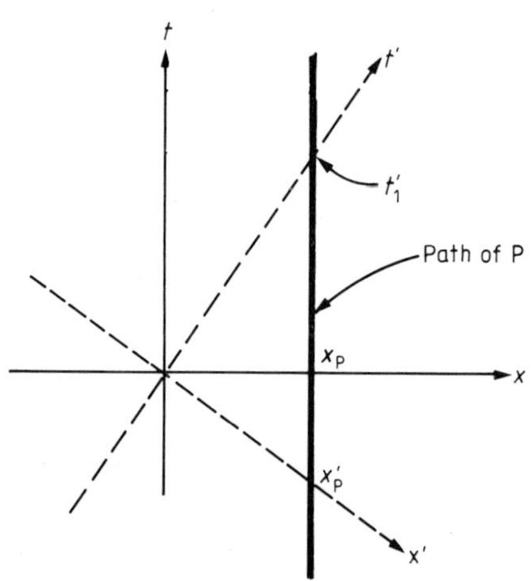

Figure 15.2 Showing the equivalence between a change of velocity of the observer and a rotation of the x,t axes. The moving observer uses the x',t' axes.

velocity relative to the second observer. This justifies the claim that changing the velocity of the observer is equivalent to rotating the axes.

As in the discussion of equation (5.3) there are quantities which are invariant under such rotations. These are the sums of the squares of the space components *minus* the square of the time component multiplied by c^2, i.e. $(x^2+y^2+z^2-c^2t^2)$. The minus sign is very important (apart from the fact that it reminds us that time is *not* a dimension of space). The factor c^2 of course makes the units correct. Note that if we only make rotations which do not involve the time direction (i.e. consider only observers without any relative velocity) then this rule is the same as that of §5.

We shall devote the next section to a particular important '4-vector' with this invariance property.

§16 Energy and momentum in special relativity

The importance of energy and momentum in physics is that they are conserved quantities, i.e. the total energy and the total momentum of a set of particles are not changed by interactions among the particles. Of course the energy and momentum have to be carefully defined so that this is true. Also, as we have noted, there are no preferred observers in the world, so all observers should see the same conservation rules. Since special relativity theory leads to modified transformation laws connecting quantities seen by different observers, the definitions of energy and momentum also have to be modified from those in pre-relativity mechanics. We shall discuss these modifications in detail; readers who do not wish to know such details should read only the last paragraph of this section.

The momentum of a particle of mass m, moving with velocity $\boldsymbol{v}$, is defined to be

$$\boldsymbol{p}=\frac{m\boldsymbol{v}}{(1-v^2/c^2)^{1/2}} \tag{16.1}$$

which differs from equation (12.2) by the square root factor. Note that this factor is nearly 1 if v^2 is much less than c^2. Like $\boldsymbol{v}$, this momentum will depend on the observer.

We now wish to define an energy of the particle in such a way that when we regard $(E, c\boldsymbol{p})$ as the (time, space) components of a 4-vector then it will have the invariance property noted at the end of the last section. To do this we put

$$E=\frac{mc^2}{(1-v^2/c^2)^{1/2}}. \tag{16.2}$$

Again, different observers will obtain different numerical values for E.

However simple algebra yields

$$E^2-c^2p^2=\frac{m^2c^4}{1-v^2/c^2}-\frac{m^2c^2v^2}{1-v^2/c^2}$$

$$=m^2c^4\left(\frac{1}{1-v^2/c^2}-\frac{v^2/c^2}{1-v^2/c^2}\right)$$

i.e.

$$E^2-c^2p^2=m^2c^4 \tag{16.3}$$

which is a constant, independent of the observer, as required. Let us make this clear: two different observers will measure the energy and momentum of a certain particle; they will obtain different results which we might call $E_1 \neq E_2$ and $\boldsymbol{p}_1 \neq \boldsymbol{p}_2$. Both, however, will agree with equation (16.3), i.e. they will find $E_1^2-c^2p_1^2=E_2^2-c^2p_2^2=m^2c^4$.

Note that, for a particle at rest (i.e. $p=0$) with regard to a particular observer, equation (16.3) (or equation (16.2)) gives

$$E_0=mc^2 \tag{16.4}$$

which is Einstein's famous equation relating energy and mass. We have put the zero suffix on E to emphasise that the equation is only true when the momentum is zero.

Let us see how these equations permit the conversion of 'mass into energy'. In fact it is mass into 'kinetic energy' that takes place, so we must first define kinetic energy (T). It is the energy due to the motion of a particle† and is therefore the difference between the energy of a particle with given momentum and the energy of that same particle at rest (equation (16.4)), i.e.

$$T=E-E_0=E-mc^2. \tag{16.5}$$

Note that T is always positive (proof: equation (16.2) shows us that E is greater than or equal to mc^2, with equality only if $v=0$). We can relate T to the expression familiar—to some readers at least—in non-relativistic physics, if we write equation (16.5) as

$$E=T+mc^2 \Rightarrow E^2=T^2+2Tmc^2+m^2c^4.$$

We compare this with equation (16.3) and obtain

$$2Tmc^2+T^2=c^2p^2.$$

† This of course is the apparent or 'useful' energy.

Now in non-relativistic physics, v is very much smaller than c, which means that T is very much smaller than mc^2, so we can neglect T^2 and find $T = p^2/2m$, which is the promised familiar result.

Now consider a decay process:

$$A \rightarrow X_1 + X_2 + X_3 + \ldots \qquad (16.6)$$

It will be convenient to view this from the position of an observer at rest relative to A, so the initial† energy–momentum 4-vector is $(M_A c^2, 0)$, where M_A is the mass of A. Let $(E_1, c\boldsymbol{p}_1)$, $(E_2, c\boldsymbol{p}_2)$, etc, be the 4-vectors associated with X_1, X_2, etc, after the event. Then energy conservation gives us

$$M_A c^2 = E_1 + E_2 + E_3 + \ldots \qquad (16.7)$$

which can be written (with the help of equation (16.5) for each particle):

$$(M_A - M_1 - M_2 - M_3 \ldots)c^2 = T_1 + T_2 + T_3 + \ldots \qquad (16.8)$$

This is a nice equation. First, since all terms on the right-hand side are positive, it tells us that A can only decay into a set of particles if its mass is greater than the sum of the masses of the particles. Second, it shows how the change in total mass is converted into the kinetic energy of the produced particles.

A composite particle is stable against decay into its constituents provided its mass is less than the sum of the constituent masses. The difference represents the energy that has been used to form the composite state and is a measure of the strength of the forces that bind the constituents together. When multiplied by c^2 it is usually called the binding energy. If the stable particle is denoted by A and the constituents by 1, 2, 3, etc, then we have

$$\text{Binding energy} = c^2(M_1 + M_2 + M_3 + \ldots) - c^2 M_A \qquad (16.9)$$

and, for a stable particle,

$$\text{Binding energy} > 0. \qquad (16.10)$$

† i.e. when we have just particle A.

§17 Strong interactions

We return to the question of the forces which hold together protons and neutrons in nuclei. By comparing the mass of a nucleus containing Z protons and N neutrons with $(ZM_{\mathrm{p}}+NM_{\mathrm{n}})$, where M_{p} and M_{n} are the proton and neutron masses, we can calculate the binding energy of the nucleus. It is much larger than atomic binding energies, so we can conclude that the forces are considerably stronger than in the atomic case, i.e. than the electric forces. These nuclear forces are thus called 'strong interactions'. Their study was the main activity in particle theory between 1950 and the mid-1970s. Many ingenious ideas were proposed, and eagerly followed, and there were several misleading 'clues'. Now we believe we have a basic understanding of these forces; it is (in principle, though not in application!) simple—as we shall see.

In contrast to the electric force (point (c) in §7), nuclear forces must be 'short-range', i.e. decrease faster than r^{-2} for particles separated by the distance r. This can be seen by noting that, although they are stronger, by a factor of about 10^3, at distances of the order of nuclear sizes (10^{-15} m), they are negligible at the atomic distances (10^{-10} m). The range can be measured directly from experiments on proton–proton scattering, and it turns out that the forces drop away very rapidly at separations greater than about 10^{-15} m.

These forces also have a more complicated structure than the electric force; in particular, they depend on the relative spin directions and velocities of the interacting particles. One simplifying feature is that they are the same for protons and neutrons, i.e. they are independent of the electric charge.

In early studies of strong interactions it was natural to regard protons and neutrons as *elementary* particles. They had spin $\frac{1}{2}$, just like the electron, and a world made of electrons, protons, neutrons and photons (electromagnetic radiation) looked attractive. However, this idea was rapidly eroded by the discovery of many more, strongly interacting, particles. These do not occur naturally (except in cosmic rays) because they decay rapidly. They are produced in experiments using accelerated particles. For example, when a proton beam falls on a proton target (e.g. hydrogen) the reaction

$$p+p \rightarrow p+\Sigma^{+}+K^{0} \tag{17.1}$$

occurs. This gives two particles, the 'sigma' and the 'K', which are new to us. Both these particles actually occur with various charges: ± 1 and 0 for the Σ, 1 and 0 for the K. In contrast to the situation with the proton and neutron we do not give the different charge states different names—there are not sufficient letters—but distinguish them by +, −, zero indices.

By measuring the momenta of the particles in the reaction (17.1) it is possible to deduce the masses of the K and Σ. It turns out that their sum is greater than the proton mass. Thus we have here an example of a reaction where the mass is *increased* (in contrast to the process considered in equation (16.6)). The extra mass comes from the kinetic energies of the original protons, so a certain minimum energy is required in order that the event can occur. This is, of course, the other reason why high energy accelerators are needed; it is not possible to make new, heavy, particles without having sufficient initial kinetic energy.

We now use the above reaction to discuss two important issues. First, does the reaction mean that a proton is made of a Σ^{+} and a K^{0} bound together by some force? It certainly looks as though one of the protons is broken up into its two constituents. (Indeed readers may recall that it was the fact that electrons could be pulled out of metals by electric fields, i.e. knocked out by photons, that told us they had to be present in atoms. A similar argument would suggest that K and Σ had to be present in a proton.) Before readers jump quickly to the answer 'yes', we should point out that there are many similar reactions that would suggest different proton constituents, such as, for instance, $\Sigma^{0}+K^{+}$. It is even possible to have a reaction like

$$p+p \rightarrow p+p+\pi^{0} \tag{17.2}$$

where the zero-charged 'pion' is another particle new to us. This seems to be saying that a p is a $p+\pi^{0}$ bound state—whatever that might mean! Thus the answer to the question has to be 'no'. It is more convenient to regard the new particles as being 'created' in the process than being there in the first place. We shall understand all this much better in the subsequent discussion, but we should mention here that models of strong interacting particles in which they are all, in some sense, composites of each other were very fashionable (and successful) during the 1960s. (Such models enjoyed the distinction of having 'nuclear democracy'—appropriate to the spirit of the '60s.)

The other issue arising from reaction (17.1) is almost the opposite of the above. If we accept that particles can be 'created' and 'destroyed' in interactions, then why should Σs, for example, only be made in the above way? Why do we not see reactions like

$$\mathrm{p}+\mathrm{p}\rightarrow\Sigma^{+}+\mathrm{K}^{+} \quad (?) \tag{17.3}$$

or

$$\mathrm{p}+\mathrm{p}\rightarrow\Sigma^{+}+\mathrm{p}. \quad (?) \tag{17.4}$$

These particular events would not require such high energy accelerators; indeed in (17.3) we have a loss of mass (i.e. $M_{\Sigma}+M_{\mathrm{K}}<2M_{\mathrm{p}}$) so the reaction could take place at zero initial kinetic energy. The simple answer is that some reactions just do not happen—not *all* reactions are 'possible'. This is an experimental fact. Fortunately, the situation is not completely chaotic and there are *rules* (which we shall later understand theoretically) which tell us what is possible. The first such rule we have already mentioned (§7(*c*)): conservation of electric charge. Note that in reactions (17.1) and (17.2) we were careful to arrange that the total charges before and after the collision were equal (+2). Charge conservation forbids a process like $\mathrm{p}+\mathrm{p}\rightarrow\mathrm{p}+\Sigma^{+}+\mathrm{K}^{+}$. No violations of this rule have *ever* been observed and there are no reasons for believing that it is not always obeyed (it is related to the zero mass of the photon and the infinite range of the electric force, as we shall see in §21).

The next rule is very similar, except that the conserved quantity is not 'measured' like electric charge but assigned to make the rule true. (This does *not* mean that the rule has no predictive power—we make the assignments from one set of processes and can then use them in others.) The quantity is called 'strangeness', and is zero for the proton, neutron and π, +1 for the Ks and −1 for the Σ. Thus the reactions (17.1), (17.2) are allowed, but (17.4) is forbidden. Later (chapter 3) we shall see that there are some rare processes that violate strangeness conservation—their story is the most intriguing of all!

A rather different conservation rule is associated with 'baryon number'. All strongly interacting particles of spin $\frac{1}{2}$, $\frac{3}{2}$, etc carry baryon number ±1, whereas particles without strong interactions and particles of spin 0 and 1 have zero baryon number (e.g. K, π).

Clearly baryon number conservation forbids the process (17.3). At present (September 1984) there is no firm experimental evidence against baryon number conservation being exact. There are, however, theoretical reasons for doubting it—we shall meet these in later chapters.

Where are we? We have lots of strongly interacting particles, with complicated interactions among them. It looks like the nuclei situation again; surely these particles are composites of a simpler set. Then the various 'reactions' are just rearrangements of the constituents, and there will be natural explanations for the conservation rules. However—unlike the situation with atoms or nuclei—there are no obvious candidates for the constituents. This is why the discovery of the truth took so much time and effort! It will not take us so long, but we shall, first, digress into some theoretical topics. We will not return to the strongly interacting particles until §22, by which time we will have scaled the highest peak of our story, from where we shall be able to see how everything fits together.

Before we leave this section, however, we comment on the difference between conventional and nuclear sources of energy. The former involve chemical reactions where the energy is derived from the electrical forces between atoms, and are therefore 'atomic' (a name which is unfortunately, and misleadingly, used for the latter) or, more generally, 'molecular'. For example, in the burning of coal, carbon monoxide molecules are formed from the carbon in the coal and from oxygen in the air. No change in the nuclei, here carbon and oxygen nuclei, is involved. The energy released can be calculated from equation (16.8) by inserting the masses of the appropriate atoms and molecules.

In nuclear sources of energy, on the other hand, the reaction changes the nuclei, e.g. as in the fission of uranium mentioned in §14. The reason why such sources are more powerful than chemical sources is that nuclear forces, and hence binding energies, which occur in the left-hand side of equation (16.8), are much greater. In consequence, more of the mass is converted into kinetic energy. The proportion of the mass that is used is, however, still quite small. This explains why antimatter has such a fascination for writers of science fiction. An antiparticle has opposite values to the particle for all the conserved quantities, so there is nothing to prevent a particle and its antiparticle from completely annihilating each other, thereby converting *all* their mass into kinetic energy of, for example, photons.

§18 Field theory

Two great general structures are at the heart of theoretical physics: quantum mechanics and special relativity. The *combination* of these structures has proved very difficult and it is not clear even now whether a satisfactory unification exists. There is, however, a framework consistent with both, and therefore called 'relativistic quantum field theory', which permits at least approximate calculations of various processes. Where the approximations are expected to be valid the results agree with experiment, but in some cases no satisfactory approximation scheme exists, and in general it is not really known whether the equations even have exact solutions. Nevertheless, relativistic quantum field theory is a fascinating subject and is crucial to the continuation of our story. It is also not easy, so we shall proceed slowly.

Understanding the motion of any mechanical system requires two stages. First, it is necessary to specify the objects involved and the forces between them. Then it is necessary to solve some equations in order to deduce what happens when the system is allowed to develop from some given starting point. In relativistic quantum field theory the first stage requires the writing down of a quantity $\mathscr{L}$, called the 'Lagrangian'. The Lagrangian *defines* the theory: it tells us what particles exist and what can happen to them. More precisely, $\mathscr{L}$ contains only those particles which are *elementary*, i.e. not made out of constituents. The non-elementary objects of the observed world should appear as solutions of the equations, i.e. as bound states of the elementary particles. Similarly, $\mathscr{L}$ only contains the *elementary* interactions; others can be built from these as described below.

We now introduce a pictorial way of representing the terms in $\mathscr{L}$. The resulting pictures, called 'Feynman diagrams', after their originator Richard Feynman, will also be used to describe the method of calculation. Let us suppose that our theory contains a set of elementary particles which we denote by A, B, C, etc. Then, since the simplest thing that can happen to a particle is nothing, we must have terms in $\mathscr{L}$ to allow the particles to move freely without interaction. This is illustrated for the free motion of A and B in figure 18.1. In this figure, and in all Feynman diagrams, no significance should be attached to the directions or lengths of the lines. Indeed, the lines should be considered as going to the edge of the picture. Thus, for example, we can think of the particle A *entering* figure 18.1 and then *leaving*, without having interacted.

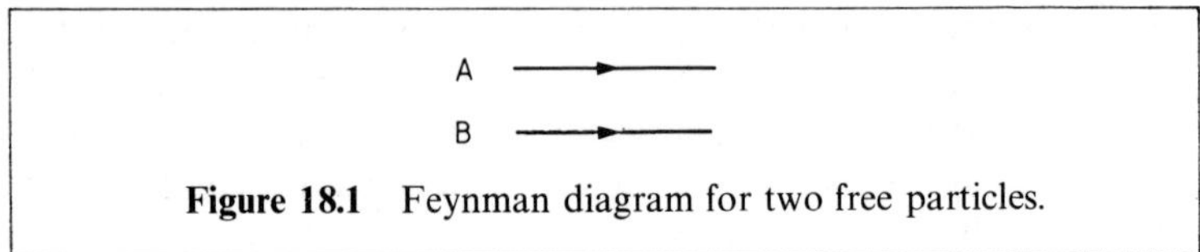

Figure 18.1 Feynman diagram for two free particles.

Non-interacting particles are not very interesting, so we include some interactions. For example, if we wish the process A→B+C to be one of our elementary interactions then it must be included in $\mathscr{L}$. Such a term is shown in figure 18.2(*a*). Here the particle A enters the diagram and is *absorbed* at the vertex where particles B and C are *created*.

We have put the letter g at the vertex to indicate that we have to specify a 'strength' of the interaction (roughly: the likelihood of A→B+C happening). This number is a free parameter of our theory

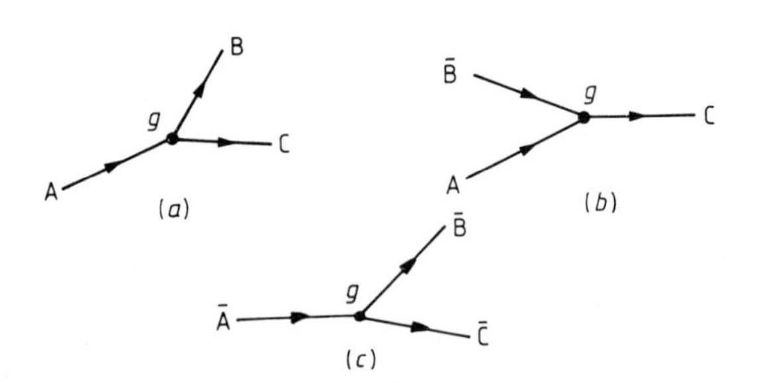

Figure 18.2 Feynman diagrams for the processes (*a*) A→B+C, (*b*) A+B̄→C and (*c*) Ā→B̄+C̄, which all correspond to the same term in $\mathscr{L}$.

and is called a 'coupling constant'. The larger g, the stronger the relevant interaction.

Now we must learn two important features of relativistic quantum field theory. First, in any Feynman diagram it is always possible to change the direction of the arrow on any line provided that the corresponding particle is replaced by its antiparticle. For example, the term in $\mathscr{L}$ which gives the diagram figure 18.2(*a*) also gives figure 18.2(*b*), which describes $\mathrm{A}+\bar{\mathrm{B}}\rightarrow\mathrm{C}$. (We remind ourselves that the orientation of the lines in a Feynman diagram is irrelevant; it is purely an aesthetic criterion that caused me to rotate the B line in figure 18.2 so that incoming particles are on the left of the interaction.) It is easy to see that this possibility is consistent with (and indeed requires) the property noted earlier that particles and their antiparticles have opposite values for electric charge and for other conserved quantities. One can also see that it explains why antiparticles *have* to exist—this argument can, in fact, be made into a proof. The second important feature is similar: we can change *all* the particles to their antiparticles. Thus figure 18.2(*c*) is also contained in $\mathscr{L}$. We see that a combination of these two features would enable us to draw many more diagrams than those shown in figure 18.2, all associated with the same term in $\mathscr{L}$. Normally we only draw one such diagram; all the others are understood to be included.

Notice that we have so far said nothing in this section about the momentum and energy relations. Whether a given process, like figure 18.2(*a*), can actually occur in nature, consistent with energy and momentum conservation, is not at present our concern. We shall return to it later.

The interaction shown in figure 18.2, which involves three particles, is of course not the only type that can be included in $\mathscr{L}$. Another example, this time involving five particles, is shown in figure 18.3. This describes the process where particles A and B collide and annihilate each other leaving three C particles: $\mathrm{A}+\mathrm{B}\rightarrow\mathrm{C}+\mathrm{C}+\mathrm{C}$. It also describes various other processes obtained by changing particles to antiparticles and reversing directions of lines, as discussed above, e.g. $\bar{\mathrm{A}}+\mathrm{C}\rightarrow\mathrm{B}+\bar{\mathrm{C}}+\bar{\mathrm{C}}$, etc.

It is clear that, even when we have chosen our set of elementary particles, the number of possible interaction terms we can include in $\mathscr{L}$ is unlimited. How do we decide what terms to include? A large part of the following story concerns the answer to this question. The Lagrangian defines the theory, but it appears to allow an unlimited

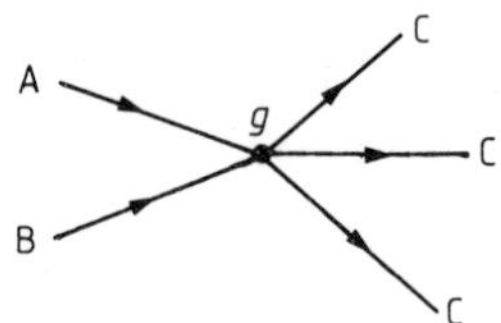

Figure 18.3 A Feynman diagram showing an interaction in which one A particle and one B particle enter and change into three C particles.

amount of freedom. Are there any principles we can use to restrict the form of $\mathscr{L}$ and, if so, can we then find a theory consistent with these restrictions and with the observed world?

Before we consider the answers to these questions we have to know something about how we calculate the rate for any particular process given the form of $\mathscr{L}$. In principle this is easy. The first step is to join together the diagrams contained in $\mathscr{L}$, *in all possible ways*, so as to make a diagram that looks like the required process. For example, if we wish to calculate the scattering process $\bar{B}+C \rightarrow B+\bar{C}$ we could use figure 18.2 and figure 18.3 once each and thereby obtain figure 18.4. The reader should check that the interactions in the circles in figure 18.4 are indeed derived from the elementary interactions of figures 18.2 and 18.3.

For the second step, we use the rules of relativistic quantum field

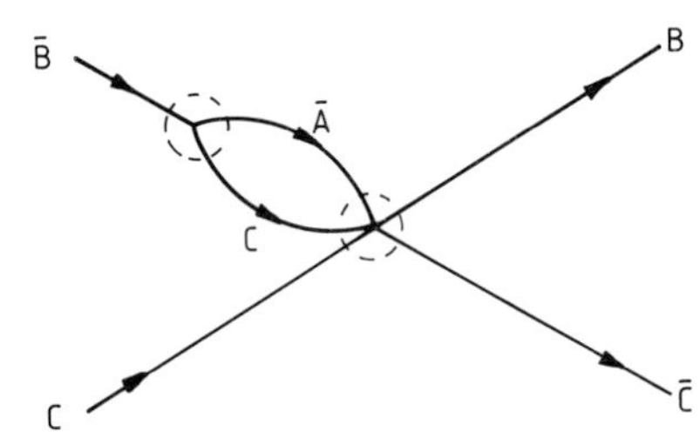

Figure 18.4 A diagram made from figures 18.2 and 18.3 which contributes to $\bar{B}+C \rightarrow B+\bar{C}$.

theory to evaluate the contribution of each diagram. Here we meet a big problem: some of these contributions turn out to be infinite! The reason for this will be explained in §20, where we shall also show how the problem can, at least in suitable cases, be avoided.

The last step of the calculation is to add the contributions of all diagrams of the required form. This introduces the second problem: there are an infinite number of possible diagrams for any process. Clearly we cannot work them all out individually. Progress is made by introducing a clever trick to add certain infinite sets of diagrams (see §20) in such a way that all but the simplest of those remaining are small enough to be neglected.

We now leave the general discussion of relativistic quantum field theory in order to concentrate on a particular example: the quantum theory of electromagnetism. This will provide an illustration of many of the points discussed in this section and will perhaps clarify them. To call it 'an' example is perhaps misleading; it is 'the' example. It successfully combines electromagnetism, quantum theory and relativity, and it certainly deserves to be treated in a separate section.

§19 QED

We consider here a world containing only electrons and photons. In the Lagrangian we put the simplest possible interaction consistent with conservation of charge†, namely that shown in figure 19.1, where we have used a dashed line, and the symbol γ (gamma), for the photon. This Lagrangian, in fact, gives the quantum theory of electromagnetism (i.e. it gives all of the electromagnetic effects discussed in the previous chapter together with quantum 'corrections' which are small in suitable circumstances). The resulting theory is called quantum electrodynamics, usually abbreviated to QED.

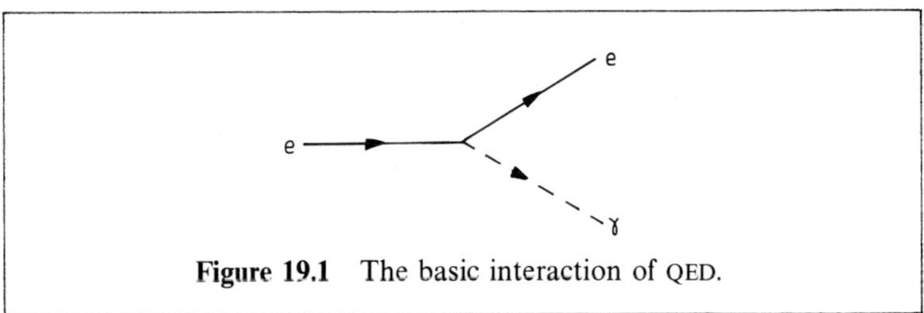

Figure 19.1 The basic interaction of QED.

Suppose we want to calculate electron–electron scattering, that is the process e+e→e+e. We might, for example, want to know the probability of an electron being scattered by another electron through an angle θ (cf the discussion of Rutherford's experiment in §9). The procedure, as we saw in the previous section, is easy to describe. We first draw all possible diagrams, involving just the interaction contained in $\mathscr{L}$, i.e. figure 19.1, in which two electron lines enter and two leave. Some examples are shown in figure 19.2. Each of these diagrams can then be evaluated by a simple set of rules. At this stage the only one of these rules we need to know is that for every

† The photon has zero charge.

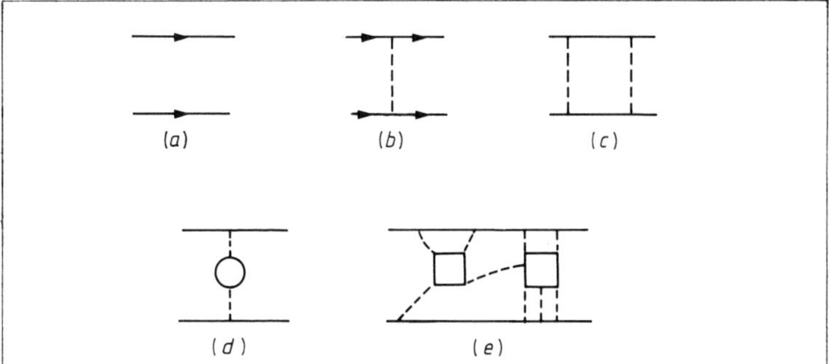

Figure 19.2 Some diagrams which contribute to electron–electron scattering.

'vertex', i.e. place where lines intersect, we have a factor g. The final answer is obtained by adding the contributions of all diagrams. It is clear, however, that this is not possible, since, as mentioned earlier, and even with such a simple $\mathscr{L}$, there are an infinite number of possible diagrams—of ever increasing complexity. We appear to be in trouble.

We shall not, however, give up, but rather look at some of the simplest diagrams. The one in figure 19.2(*a*) shows the two electrons not interacting. This does not contribute to the probability of scattering through any non-zero angle, and is therefore not of interest. Diagram (*b*) is the simplest interaction. It can be thought of as the exchange of a photon between the two electrons. Since it has two vertices it yields a contribution proportional to g^2. Explicit calculation shows that this contribution is exactly equal to the interaction introduced in §7 provided that we identify the coupling constant g with the electron charge e. What about the other diagrams? These represent more complicated contributions to electron scattering; in (*c*) there are two photons exchanged whilst in (*d*) the photon which is exchanged splits into an $e+\bar{e}$ pair which then recombine; clearly other diagrams can similarly be described, though such descriptions are not really helpful. Of more significance is the fact that they all contain more than two vertices and hence more than two factors of g: (*c*) and (*d*) contain g^4, (e) contains g^{18}. The power of g in a given diagram is called its *order*, so these are all *higher-order* diagrams. Now suppose g is very small, then every extra factor of g

reduces the contribution of a diagram and it may then be a reasonable approximation to ignore the contributions of all but the lowest order diagram(s), thereby making an apparently hopeless task very simple. This is the key to almost all the results obtained from relativistic quantum field theory.

In the present context we must now ask whether the actual value of g ($\equiv e$) in QED is in fact sufficiently small. Since this number requires units (see §7) we must first find an appropriate 'scale' so that we can form a number in which the units cancel (whether or not a number is 'small', in the sense being considered here, should not depend on the units in which it is expressed). The only other numbers with units that enter our problem are h and c, so the appropriate quantity is e^2 divided by the product of h and c. In fact, the quantity usually considered is

$$2\pi e^2/hc \simeq 1/137. \tag{19.1}$$

Fortunately this is a small number, so the approximation of keeping only the lowest-order term is reasonable. It is this fact that explains why we were able to use classical electromagnetism, in particular the electric force, with such success in the previous chapter.

There are, of course, corrections to the first-order results. These can be evaluated approximately by calculating the next few orders. The effects they produce have been measured in extremely accurate experiments, and the agreement between the results of such experiments and the theoretical predictions provides impressive evidence for the validity of QED.

Before we leave this section we return to the strong interaction, which is the main topic of this chapter, and note that a Lagrangian to describe this would have to differ in at least two important respects from the QED Lagrangian. First, the coupling constant would be greater than 1—so there would be no justification for keeping only the lowest-order contribution, and hence no way of calculating any result.

Second, the particles playing the role of the photon would have a mass, and the effect of this would be to give the force a short range. To understand this we note that a process like that involved at the top of figure 19.2(*b*) cannot satisfy conservation of energy and momentum when the dashed line is associated with a massive particle. It is, in fact, an example of the decay process explained at the end of §16; clearly the final state mass is greater than the initial mass. Thus the process cannot happen. However, we here remember the 'uncertainty principle', which is properly built into the equations of quantum mechanics (see §12). For small distances, 'errors' are allowed in the momenta (similar 'errors' are allowed in the energy for small times). In actual calculations the energy and momentum conservation rules are enforced at each vertex, but the relation defining the mass, equation (16.3), is only correct for particles which are free to travel over large distances, e.g. those on the external legs of the diagram. Where a particle exists only for a short time or distance then deviations from the mass relation can allow otherwise forbidden processes to occur. This of course will be the situation in diagrams like figure 19.2(*b*) *provided* that whatever absorbs the massive particle (i.e. the bottom part of the diagram) is sufficiently close to the particle that emits it. Thus the exchange can only take place over small distances. Clearly, then, the *heavier* the exchanged particle the *shorter* is the range of the corresponding force. The extreme case is the photon, which is massless and gives a force of infinite range (recall the discussion of §7). The range of strong interactions (10^{-15} m) corresponds to a mass of about one tenth of the proton mass, i.e. about the mass of the pion introduced in the previous section.

In the next section we shall see that there are some problems with what we have been doing here. Remarkably, we shall be able to use these problems to tell us exciting new things!

§ 20 Renormalisation

As we have discussed in the previous section the calculation of a given process requires adding together an infinite number of contributions. In practice, it is possible to obtain answers only if the coupling constant(s) is (are) sufficiently small so that the total contribution of all diagrams, except a few simple ones, is negligible. Of course, even when the coupling constant is small, so that individual high-order diagrams are negligible, it is not obvious that their sum will be small. The number of diagrams increases with the order and we have to sum over an infinite number of orders. This type of situation is well known in mathematics, and there are many theorems. Unfortunately, none are suitable to tell us whether our infinite sum really does 'converge' to a finite answer. We rely here on hope, and on the success of the predictions.

In addition to being concerned about the number of diagrams of a given order, we must think about the size of the contribution of a given diagram. So far we have only said that it contains a factor of g^n, where n is the order. What other factors does it contain? The precise details again need not concern us, but as we noted in §18 there is one factor that in some circumstances can be infinite. This we cannot ignore!

Let us first see how the infinity arises. To do this we must think about momentum conservation at the various interactions. In figure 20.1 we show again the simplest electron scattering diagram, figure 19.2(*b*), this time with the momenta of the particles indicated. Initially we have two electrons with momenta $\boldsymbol{p}_1$ and $\boldsymbol{p}_2$; after the interaction we have momenta $\boldsymbol{p}_1'$ and $\boldsymbol{p}_2'$. At each interaction momentum is conserved, i.e.

$$\boldsymbol{p}_1 = \boldsymbol{p}_1' + \boldsymbol{q} \tag{20.1}$$

and

$$\boldsymbol{p}_2 + \boldsymbol{q} = \boldsymbol{p}_2'. \tag{20.2}$$

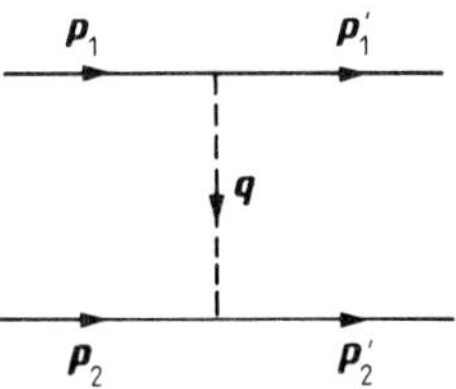

Figure 20.1 The simplest diagram contributing to electron–electron scattering, with the particle momenta given.

Consistency of these two equations requires that

$$p_1 + p_2 = p_1' + p_2'. \tag{20.3}$$

This is momentum conservation for the complete process. The point to notice here is that, once the external momenta (the $\boldsymbol{p}$) are fixed, there is no further freedom: all other momenta (in this case just $\boldsymbol{q}$) are determined.

Now we consider a particular higher-order diagram for the same process: figure 20.2. Here we have 'solved' the momentum conservation equations at each interaction, so there are no redundant momentum labels. We note, however, that there is now some freedom because $\boldsymbol{k}$ is not determined by the external momenta—it can take any value. The contribution of this diagram is therefore obtained by adding the contributions for all possible values of $\boldsymbol{k}$. (Actually, since $\boldsymbol{k}$

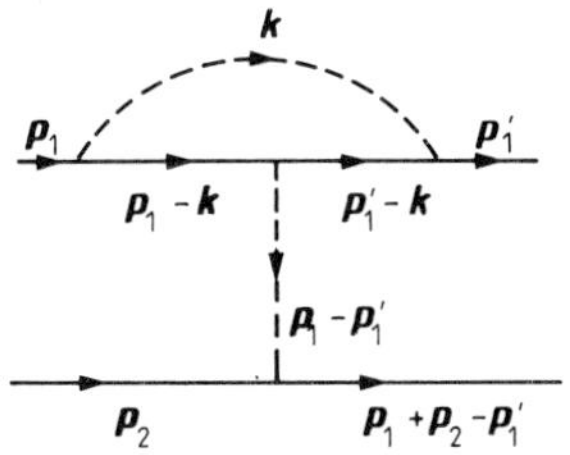

Figure 20.2 Another diagram contributing to electron–electron scattering. Again the particle momenta are indicated and it can be seen that there is one undetermined internal momentum ($\boldsymbol{k}$).

takes a continuous range of values, this is a special form of addition known as 'integration', but this is a detail which does not affect the discussion.) This is, of course, an infinite sum because there is no limit to how large $\boldsymbol{k}$ can be. In some cases such infinite sums give a finite answer. In others the terms grow with $\boldsymbol{k}$ and the sum is infinite.

At first sight this appears to be a disaster for the basic ideas of quantum field theory. Nature, however, has a marvellous habit of turning apparent disasters into beautiful theories†.

The basic idea is a little subtle, although it can be made quite precise and rigorous. We note, first, that the diagram in figure 20.2 can be 'included' in figure 20.1 if we write the top interaction as shown in figure 20.3. Here the shaded blob includes both contributions and, as indicated, an infinite set of similar diagrams. The difference between

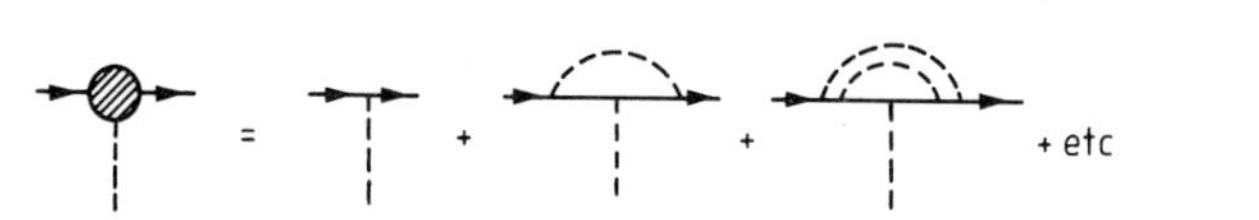

Figure 20.3 Showing how a set of diagrams can be added to renormalise the coupling constant.

the shaded blob and the first term is that it corresponds to a different value for g, the coupling constant. Thus we can *either* use the value of g in our original Lagrangian and include all terms in figure 20.3 *or*, more simply, we can use a new value for g (the 'renormalised' value) and just include the simple lowest-order interaction; the results are identical. Now comes the crucial point of the argument: we do not *know* the value of g in the original $\mathscr{L}$—it is an input parameter—so the fact that we cannot calculate the extra contributions in order to find the renormalised g is not a problem; we merely let the renormalised value of g be the input parameter. It is *this*, not the original value, that is the 'observed' coupling constant. Note that nowhere in this argument does it matter that the *difference* between the original and renormalised couplings is infinite; we just suppose the original value is also infinite in such a way that the renormalised

† A cynic might say this is just theoretical physicists being cunning and devious. Don't believe him!

value is finite and agrees with the observed value. (The mathematically sensitive will of course worry about this playing around with infinite numbers—they can read bigger books than this where, hopefully, they will be given arguments that will satisfy them. Those whose aesthetic sensitivities are offended have a greater problem—and they are in good company. The success in calculations of the higher order terms of QED, mentioned earlier, gives assurance however that, at least, the *answers* are correct. Maybe when we really include everything properly the infinities will not be there—see §33.)

Having got rid of one series of infinities we have to ask whether there are others, i.e. are there other diagrams not included in figure 20.3 which are infinite? In general the answer is yes. In particular, it is usually necessary to make an (infinite) renormalisation of the masses which appear in the original Lagrangian. Indeed, for most Lagrangians it is impossible to remove all the infinities without introducing an ever increasing number of new renormalised coupling constants. The theory thus requires an infinite number of input parameters and is therefore useless, since it has no predictive power. Such theories are said to be 'non-renormalisable'.

Now comes the exciting part! Recall that we seemed to have too much freedom in choosing our Lagrangian. We are pleased therefore to have found a criterion which restricts this freedom: we require that $\mathscr{L}$ be *renormalisable*. It turns out that this is very useful; it appears to rule out all but a small, and very beautiful, class of theories. To these we now turn.

§ 21 Gauge theories

Gauge theories have been the inspiration behind almost all that has happened in the theory of elementary particles within the last 15 years. The basic idea is very beautiful and really quite simple. In this section we shall first describe this idea in a very general way and then we shall present the technical details in the context of a particular example.

We recall, from §12, that the probability of finding a particle at any point is related to a wave-like function which depends on the space point and on the time. The Lagrangian can be written as a mathematical expression involving these functions, one for each particle type in our theory. We start with the simplest possible Lagrangian, namely the one in which there are no interactions. It therefore just contains the terms corresponding to figure 18.1.

To make a gauge theory we *notice* two things and then *do* one thing. We first notice that our Lagrangian does not change when we make particular changes to the functions describing the particles, i.e. it is *invariant* under such changes. This is very much analogous to the fact that the length of a line does not alter when we change the axes used to describe it (see §5).

The second thing we notice is that the changes which we are allowed to make to the functions describing the particles must be done in an identical way at all points of space and time. What this means is that, although there is some freedom in the functions we use to describe the particles, when we have chosen a particular description at one point of space and time, *there is no further freedom left*; observers at other times or spatial points are *not* free to choose their own way of describing the particles. We express this by saying that the Lagrangian has a 'global' invariance, the global implying that identical changes must be made everywhere and at all times.

Now we have to *do* something. We modify the Lagrangian so that the restriction to only global changes is removed. By adding suitable terms to our $\mathscr{L}$, we form a new Lagrangian which is unchanged when the descriptions used for the particles are varied *independently* at all points of space and time. Such a Lagrangian is then said to have 'local' invariance. (Note that local invariance implies global invariance but global invariance does not imply local invariance.)

It is not obvious, but turns out to be the case, that it is possible to make a locally invariant $\mathscr{L}$ and that the method of doing this is essentially unique. The locally invariant $\mathscr{L}$ has exciting properties: it contains some new functions, corresponding to massless particles of spin 1, which interact with the original particles. Thus, from our theory without any interactions, we have built a very specific theory with interactions. All the excessive freedom in adding different types of interaction (noted in §18) has disappeared.

What has all this to do with renormalisability? The answer has probably been guessed. *The locally gauge invariant theory is renormalisable.* Essentially this is because it is very tightly constrained, in such a way that potentially infinite diagrams cancel among themselves.

Such a beautiful idea just has to be relevant to the real world. Indeed it is: in the simplest example of the theory we have described the massless particle is the photon and the Lagrangian obtained is that of QED. We should pause here to note that this is really rather amazing. A particular property of the function describing free particles is unobservable (see below for more details). By making the global invariance, implied by this statement, into a local invariance, we have *derived* all of electromagnetism which, as we recall from chapter 1, is the major part of observable physics. We have also, as promised, understood the remark regarding electromagnetism made in §3. Of course, historically, things did not actually happen like this. The laws of electromagnetism were obtained by careful analysis of laboratory experiments, and it was only later realised that local gauge invariance was a property of these laws.

There is no doubt that the idea of local gauge invariance is so elegant—and so productive of correct results—that we would like to use it again. Can it, or something similar, help us with our problem of the strong interaction? There are some difficulties here. What *other* global invariance exists that we might try to make into a local invariance, thereby introducing new interactions? How can we obtain forces of short range when the particles we introduce in the interaction have zero mass? We shall answer these questions.

First, however we give a more technical description of QED as a gauge theory. Readers who find this too difficult should jump immediately to the next section.

We begin with the wave function which, according to quantum mechanics, is associated with each particle type. This function, which we shall here denote by ψ, depends upon position ($\boldsymbol{x}$) and time (t). It is not, as we have previously implied, just a number; rather it should be thought of as a line in a two-dimensional space, as shown in figure 21.1. As many readers will know, such quantities are normally called

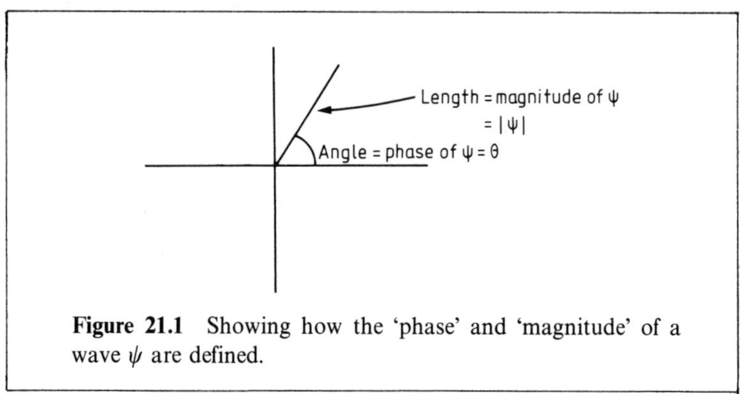

Figure 21.1 Showing how the 'phase' and 'magnitude' of a wave ψ are defined.

complex numbers. The magnitude of ψ is the length of the line, usually denoted by $|\psi|$. Its phase, θ, is the angle the line makes with some arbitrary fixed line. According to quantum theory the probability of finding the particle at a given point is proportional to the square of the magnitude of ψ at that point. What then is the meaning of the phase? The answer is that it has no meaning whatever: it is unobservable. This non-observability arises from the fact that observable physics always depends on the product $\bar{\psi}\psi$, where $\bar{\psi}$ is defined to be the same as ψ but with the opposite sign for the phase, and where the rules for taking products of these (complex) numbers are such that the phases are added. Thus the phase of $\bar{\psi}\psi$ is always zero, i.e. $-\theta+\theta$, where θ is the phase of ψ.

At this stage readers are wondering why we confuse the issue by introducing phases into the discussion, since we are never going to see them. One answer is that *relative* phases, e.g. of waves arising from different sources, *are* observable; they affect the way in which the waves add together, so interference is now a rather more complicated topic than for the water waves discussed in §11.

Next, we must consider the mathematical expression for $\mathscr{L}$. The term corresponding to the non-interacting particles contains the product $\bar{\psi}\psi$ which, as noted above, is independent of the phase. To understand the relation of this with the appropriate diagram (figure 18.1) we should associate ψ with the incoming particle which is annihilated and $\bar{\psi}$ with the created outgoing particle. Alternatively, if we recall that reversing an arrow changes a particle to its antiparticle, we can regard ψ as creating an antiparticle and $\bar{\psi}$ as annihilating an antiparticle.

The $\bar{\psi}\psi$ term in $\mathscr{L}$ is in fact the energy of a free particle and, as such, it contains a contribution from the mass, namely $m\bar{\psi}\psi$, where m is the particle mass and $\bar{\psi}\psi$ is the probability for the particle to be present, together with a contribution from the kinetic energy (cf equation (16.5)). We cannot here derive the latter contribution, but we can make it plausible in the following manner. The kinetic energy is related to the momentum, which in turn is related to the wavelength ($p = h/\lambda$). Clearly, high energy thereby corresponds to high p or low λ, i.e. to a rapidly varying ψ. Thus the kinetic energy must contain a factor measuring how rapidly ψ varies, in other words something like the difference between ψ at two nearby points divided by the distance between the points. To make this more precise in a simple way, we suppose that there is only one space direction, x, so that ψ depends just on x and on time. We sometimes make this dependence explicit by writing $\psi(x, t)$, which means the (complex) number ψ at the particular position x and at the time t. Then we require the quantity defined by

$$\frac{\psi(x+\varepsilon, t) - \psi(x, t)}{\varepsilon} = \frac{\partial}{\partial x}\psi(x, t). \tag{21.1}$$

Actually, we should calculate this quantity for ε sufficiently small so that it no longer depends on ε. There are suitable theorems in mathematics which allow this. As many readers will know we have thereby defined the *derivative* of ψ with respect to x.

The required free-particle term in $\mathscr{L}$ is thus something like:

$$\mathscr{L}_0 = m\bar{\psi}\psi + \bar{\psi}\frac{\partial}{\partial x}\psi + \ldots \tag{21.2}$$

There are additional pieces associated with other dimensions of space, and one associated with the time variation of ψ, but equation (21.2) is adequate for our purpose. (Note that we are ignoring numerical factors when writing equations like (21.2) since they are not relevant to our purpose. Readers who are—quite rightly!—concerned about units will realise that the first term must be multiplied by c/h to maintain consistency.)

We are now in a position to repeat the steps discussed earlier in this section. First, recalling that the phase θ disappears from $\bar{\psi}\psi$, we *notice* that $\mathscr{L}$ in equation (21.2) is not changed if we alter θ. This is an utterly trivial point which can be expressed in rather grandiose language:

'The	free	Lagrangian has	global	U(1) gauge invariance.'
	↓		↓	↓
	means no inter-actions		means that the phase change is a constant—it does *not* depend on space or time	means we change the phase—don't trouble to ask why

Next, we *notice* the important word 'global'. It is easy to see why it is necessary. If we had introduced, for example, a different change of phase at x and $x+\varepsilon$, then it would not have been cancelled when we evaluated the product $\bar{\psi}(x, t)\psi(x+\varepsilon, t)$. The $\mathscr{L}_0$ would therefore have changed, i.e. it would not be invariant. (For those with the necessary mathematics this can be easily calculated. We consider the phase change $\psi \rightarrow e^{i\lambda(x)}\psi$, then $\bar{\psi} \rightarrow e^{-i\lambda(x)}\bar{\psi}$ and $\bar{\psi}(\partial/\partial x)\psi \rightarrow \bar{\psi}(\partial/\partial x)\psi + i\bar{\psi}\psi(\partial\lambda/\partial x)$. Thus $\mathscr{L}_0$ has changed by the addition of $i\bar{\psi}\psi(\partial\lambda/\partial x)$.)

So, the phase is unobservable but, once it is fixed at one point of space and time, it is fixed for every point. We are not all free to choose our own phase. There is nothing *obviously* wrong with this but it might be considered aesthetically unsatisfactory and, in any case, it seems that nature does not like it. Let us then see if we can modify $\mathscr{L}_0$, so that it is unchanged when we make a change of phase which is an arbitrary function of space and time. There is a unique way of doing this. It is necessary to introduce another function, corresponding to a spin 1 particle, which, under the arbitrary phase change, itself changes in such a way so as to cancel the change in $\mathscr{L}_0$. (Explicitly, we add $g\bar{\psi}\psi\phi$, where g is an arbitrary constant, and assume that, under the phase change above, $\phi \rightarrow \phi - (i/g)(\partial\lambda/\partial x)$.)

So, where are we? We began with the Lagrangian of a free particle, say, an electron. By demanding local invariance we have constructed a new Lagrangian which has, in addition, a spin 1 particle and an interaction term. The new particle is the photon, the interaction is that of figure 19.1 and, as promised, we have derived the Lagrangian of QED. We emphasise in particular that the zero mass of the photon and the property of conservation of charge are derived consequences. Local gauge invariance just does not permit either a photon mass or terms in $\mathscr{L}$ that would change the total charge.

Since the end of §17 our discussion has been rather formal and it is time we returned to the elementary particles, in particular to the strong interactions, and learned how these formal ideas provide the answer to almost all our problems. Readers who have been finding recent sections a little too 'theoretical' will enjoy meeting quarks in the next section.

§22 Quarks

The theory of electromagnetism is elegant, simple and powerful, explaining atoms and their properties. On the other hand the strong-interaction part of physics ('nuclear' physics) has a complicated pattern of short-range forces; it contains many particles, which transform into each other, and there are no clear indications as to which should be regarded as elementary and which as composite. The key to the understanding of this mystery, which was the principal concern of particle physics throughout the 1950s and 1960s, came with the realisation that the strongly interacting particles are all composites of a small number of *new* spin $\frac{1}{2}$ particles. This was first suggested, independently, by Gell-Mann and by Zweig in 1964. The particles are called *quarks*, a name which Gell-Mann found in Joyce's novel *Finnegans Wake.*

We now know that there exist at least six types of quark. These are shown, with their electric charges and their masses, in table 22.1. This table includes the 'top' quark for which there is only very recent experimental evidence but which has for some time been expected to exist in order to complete the pattern (see §27 for further discussion). The names used for the quarks are not intended to be indicative of their properties and have only historical significance.

Table 22.1 The known quarks.

Name	Symbol	Charge	Mass (with $M_p=1$)
Up	u	$\frac{2}{3}$	5×10^{-3}
Down	d	$-\frac{1}{3}$	10×10^{-3}
Charmed	c	$\frac{2}{3}$	1.5
Strange	s	$-\frac{1}{3}$	0.15
Top	t	$\frac{2}{3}$	about 40
Bottom	b	$-\frac{1}{3}$	5

The strongly interacting particles we have already met can be made from three quarks (these are the *baryons*) or from a quark–antiquark pair (these are *mesons*). We show their quark structures, along with those of some other states in table 22.2. Clearly a large number of other baryon and meson states can also be made.

Table 22.2 The quark content of some of the strongly interacting particles.

Name	Symbol	Quark structure	Spin	Mass (approx)
Proton	p	uud	$\frac{1}{2}$	1
Neutron	n	udd	$\frac{1}{2}$	1
Delta	Δ^{++}	uuu	$\frac{3}{2}$	1.23
	Δ^{+}	uud	$\frac{3}{2}$	1.23
	Δ^{0}	udd	$\frac{3}{2}$	1.23
	Δ^{-}	ddd	$\frac{3}{2}$	1.23
Sigma	Σ^{+}	uus	$\frac{1}{2}$	1.19
	Σ^{0}	uds	$\frac{1}{2}$	1.19
	Σ^{-}	dds	$\frac{1}{2}$	1.19
Pion	π^{+}	$u\bar{d}$	0	0.14
	π^{0}	$u\bar{u}+d\bar{d}$	0	0.14
	π^{-}	$\bar{u}d$	0	0.14
Kaon	K^{+}	$u\bar{s}$	0	0.5
	K^{0}	$d\bar{s}$	0	0.5
	$K^{-}=\overline{K^{+}}$	$\bar{u}s$	0	0.5
	$\overline{K^{0}}$	$\bar{d}s$	0	0.5
Psi	Ψ	$c\bar{c}$	1	3.1
Upsilon	Υ	$b\bar{b}$	1	9.5

We see from this table that normal matter, which contains protons and neutrons, only requires two types of quark, u and d. At the time when the quark model was first introduced physicists knew of the existence of some particles (which had been referred to as 'strange') that could naturally be understood as containing the third quark (s). Thus, the original quark model only had the three lightest quarks.

A few other interesting features of table 22.2 should be mentioned. The uud states of the proton and the Δ^{+} differ because in the latter the three spin $\frac{1}{2}$ *add* to produce a total spin of $\frac{3}{2}$, whereas in the former two of them cancel to produce the spin $\frac{1}{2}$ of the proton. The π^0 state is a combination of $u\bar{u}$ and $d\bar{d}$ states; this means that either of these pairs is equally likely to be inside a π^0. This state has the property of being its own antiparticle, i.e. $\overline{\pi^0} = \pi^0$. Although not revealed in the table, different charge states do have slightly different masses. This is partly due to the different masses of the u and d quarks and partly to electromagnetic forces between the quarks.

All the interactions which we discussed in §17 can now easily be seen to be rearrangements of the quarks; no quark is required to change into another, they merely combine in different ways. Thus, for example, the process (17.1) takes place as shown in figure 22.1. Note

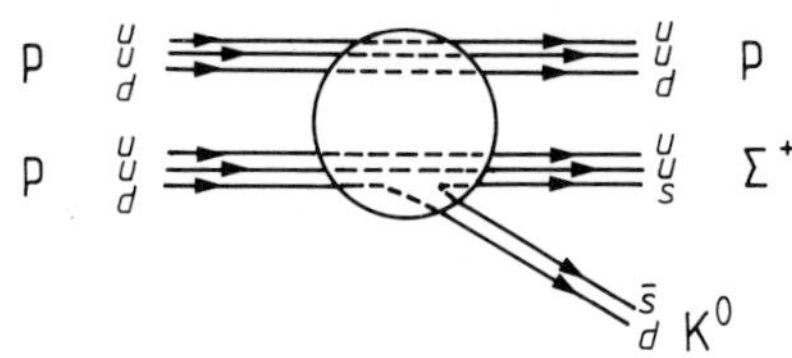

Figure 22.1 Showing how the process $p+p \rightarrow p+\Sigma^+ + K^0$ can occur through a 'rearrangement' of quarks.

that this is not a field theory diagram of the type introduced in §18 sincc at present we have said nothing about the interactions of quarks, so we cannot *calculate* anything. We are ignorant of what is happening inside the circle. Diagrams like this, however, tell us what processes are possible and, in particular, show us how we can understand conservation rules. For example we could not draw a similar diagram for the process $p+p \rightarrow p+\Sigma^+$, which would violate strangeness, without having a d quark change into an s quark. (The reader should try.) Thus, provided the forces are such that this is not possible, we will have explained the conservation of strangeness, i.e. the net number† of s quarks does not change. Similarly of course we will obtain conservation of 'charm', of 'bottom' and of 'top'.

A feature of figure 22.1 which is worth noting is the existence of 'pair production', i.e. an $s\bar{s}$ pair emerges out of nothing. Such a process is

† By this we mean the number of s quarks *minus* the number of $\bar{s}$ quarks.

possible for any particle and antiparticle pair, since they have zero value of any conserved quantity. Notice that we could draw the same picture as figure 22.1 for other processes, related by changing the directions of arrows, and, in general, different pair productions or pair annihilations will be involved. For example, the process $\bar{p}+\overline{\Sigma^{+}}\rightarrow \bar{p}+\bar{p}+K^{0}$ involves production of a $d\bar{d}$ pair rather than an $s\bar{s}$ pair. (Again readers are recommended to try drawing the appropriate diagrams.)

At first the quark model was not taken very seriously (even by its initiators!). At the time, it was generally believed that complicated mathematical structures (group representations) were at the heart of strong interaction physics, and the quark model was indeed regarded as simply a convenient way of doing the algebra of group theory. (Why did we think the truth was so complicated?) Slowly the situation changed and the model is now universally accepted. We will describe some of the reasons for this.

(i) Throughout the 1960s strongly interacting particles were continually being discovered. They all fitted beautifully into the patterns expected on the basis of 3-quark or quark-plus-antiquark states, and could not be explained simply by 'group theory'. Real quarks were needed.

(ii) It was possible to use very simple quark model arguments to compare the rates for various interactions, and these agreed well with experiment. The simplest example of this is the comparison of the proton–proton 'total cross section' (i.e. likelihood of any sort of collision) with the corresponding quantity for, say, π–proton collisions. In the first case we have 3 quarks hitting 3 quarks, i.e. 9 possible collisions, whereas in the second we have 2 hitting 3, i.e. 6 collisions. Thus we predict, correctly, a ratio of $9/6=3/2$ for the cross sections.

(iii) In a remarkable repetition of history, the existence of small charged objects within protons was observed in high-energy scattering of electrons. This is analogous to Rutherford's discovery of the small nucleus inside an atom (§9). The scale is somewhat different of course; here the proton was known to have a size of about 10^{-15} m and the charged objects were observed to be small compared with this. It is an interesting comment on the reluctance of theorists to accept the quark model that Feynman did not call these small objects quarks but invented a new name for them; he called them 'partons'. (Actually some of the partons are not quarks but 'gluons', which we meet later.)

(iv) In 1975 particles containing the charmed quark ($c\bar{c}$ states) were discovered, e.g. the Ψ in the table. The existence of such a quark had been suggested (essentially to form a partner to the *s* quark, as we shall explain in the next chapter), and the discovery of these particles gave great confidence in the whole quark model idea. In particular, because the mass of states containing *c* quarks is approximately equal to the sum of the quark masses, it is possible to do more realistic calculations than are possible in the non-charmed sector. Agreement with observation is excellent. Later, objects containing *b* quarks, e.g. the Υ, were found.

Of course, there were problems with the model. (Otherwise it would have been 'believed' much sooner, physicists being not entirely dumb.) The obvious one is that nobody was able to find any free quarks! They would be easy to identify since they have non-integral charge (in units of the electron charge). People looked—in all sorts of places—but did not find such objects. (More recently there have been some claims to see evidence for non-integral charges, but these claims have not been confirmed.) A related problem is that some quark combinations, e.g. qq, $\bar{q}qq$, $qqqq$, etc, are not seen.

The other difficulty we discuss is more technical. We recall that for spin $\frac{1}{2}$ particles a state has to be *antisymmetrical.* However the observed 3-quark states are, apparently, symmetrical. We can see this most easily for the Δ^{++} particle which is a *uuu* state of charge $+2$. The spin is $\frac{3}{2}$, so the three *u* quarks have their spins pointing in the same direction. We therefore have three quarks in identical states (this is also true for their average positions in space), hence interchanging them does nothing, i.e. the state is *symmetrical* and therefore should not exist. Many attempts were made to overcome this problem. The simplest proved correct. It also suggested the solution to the previously noted problem of non-appearance of certain states and, indeed, to the problem of the origin of strong interactions.

§ 23 Colour and QCD

The quark model state for the known baryons is, as we have seen, symmetrical in the quarks, whereas very general principles tell us that it must be antisymmetrical. The 'obvious' solution to this dilemma is to invent a new extra label for the quarks, i.e. to imagine that each quark comes in several forms, and arrange that the observed states are antisymmetric in this label. Since we do not see any evidence of the new label in the observed states (there is only one kind of proton, or neutron, for example) we want to make sure that there is only *one* 3-quark state which is antisymmetric in the new label. Hence, after a bit of thought, we realise that it must take three values. We therefore suppose that each of the previously introduced quarks can be red, blue or green†. If we then write the colour of the first quark in the first position, etc, we can have a state of three quarks, e.g. RBG. This can be made antisymmetrical in the three quarks in a unique way:

$$\mathrm{RBG}+\mathrm{BGR}+\mathrm{GRB}-\mathrm{RGB}-\mathrm{GBR}-\mathrm{BRG}.$$

(Check that if we make an interchange of any pair of colours then this expression changes sign.) There is no other 3-quark colour state which is antisymmetrical.

We can now introduce a 'rule' which allows us to predict, correctly, which quark states will be observed as free particles: such states must be *colourless*. A precise definition of this concept requires some knowledge of group theory (colourless states are singlets of the SU(3) group), but it is easily understood as the colour analogue of a state of zero electric charge (a neutral state). Here the situation is more complicated because, whereas there is only one sort of electric charge, we have three colours. When a colourless state of a given number of

† The use of these somewhat frivolous names is probably explained by the fact that, when they were introduced, the idea was almost a joke. It seemed to make an unlikely story even more implausible; it was a 'fiddle' to solve a problem with a model that nobody really believed!

quarks and antiquarks exists, it will carry no colour labels. Thus the rule obviously does not allow free quarks (which come in *three* colours) and, less obviously, eliminates all qq, $qq\bar{q}$ and other unwanted states. Notice that at this stage we do not have any explanation for the rule; it is no more than a convenient way of describing the observed facts. However, we are now in a position to make the crucial remark that does explain the rule and that leads us to the solution of strong interaction physics!

First, recall that at the end of §21 we speculated on whether there might be other global symmetries of a free Lagrangian that we could make into local symmetries, thereby producing new interactions. *The introduction of 'colour' provides us with just such a symmetry*; the Lagrangian is invariant under any mixing of the colour labels. To be explicit, the free Lagrangian, for each type of quark (u, d, etc), has the form

$$\mathscr{L}_0 = \text{`}\bar{\psi}_R\psi_R + \bar{\psi}_B\psi_B + \bar{\psi}_G\psi_G\text{'} \tag{23.1}$$

where the suffixes refer to the colours. We have put this expression in quotes because we have ignored other features of $\mathscr{L}_0$ which are not relevant to us here. This expression looks like that for the length of a line given in equation (5.3) $(x^2+y^2+z^2)$ and we recall that such an expression is invariant under rotations of the axes, i.e. under suitable mixing of x, y and z. In the same way $\mathscr{L}_0$ is invariant under mixings of the colours. (It is easy to give a simple example. Suppose we make the 'rotation in colour space': $\psi_R \to a\psi_R + b\psi_B$, $\psi_B \to -b\psi_R + a\psi_B$, $\psi_G \to \psi_G$, where a, b are (real) numbers. Then simple algebra shows that $\mathscr{L}_0 \to (a^2+b^2)(\bar{\psi}_R\psi_R + \bar{\psi}_B\psi_B) + \bar{\psi}_G\psi_G$. This equals $\mathscr{L}_0$ provided we choose $a^2+b^2=1$.)

Now the proper form of $\mathscr{L}_0$ involves derivatives (equation (21.2)), so it is clearly necessary here that the mixing is the same at all points of space and time, i.e. that a and b used above are constants. As in §21, to proceed we want to add terms to $\mathscr{L}_0$ so that this restriction can be dropped, i.e. we want to change the global invariance into a local invariance. Again this requires the introduction of massless spin 1 particles coupled to the quarks, as indicated in figure 23.1, where the broken lines denote the new particles. These particles are responsible for holding quarks together and are therefore called 'gluons'. Note that, because the invariance involves mixing the quarks, the gluons can change the colour of a quark, i.e. diagrams like those in figure 23.1 (*b*) and (*c*) arise. There would appear at first sight to be 9 gluons (3×3)

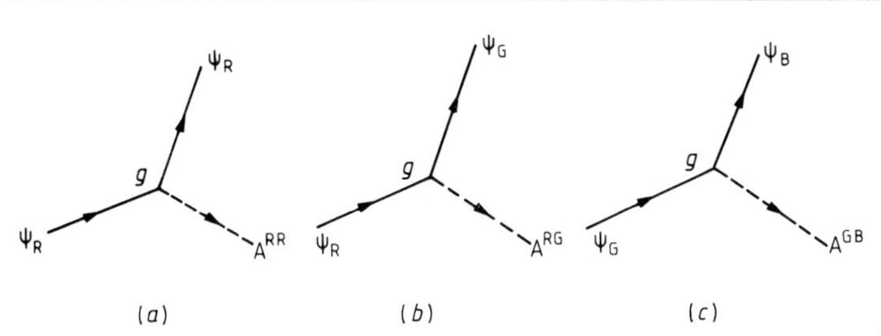

Figure 23.1 Some of the interactions that are introduced between quarks and gluons when the colour mixing symmetry is made local.

but in fact the combination (RR + GG + BB) actually corresponds to the phase change which has already been discussed and which gives QED. Thus we have 8 gluons in our Lagrangian.

Since the gluons carry colour labels, e.g. RG, GB, etc, they interact among themselves. In fact, the locally gauge invariant Lagrangian must also contain interaction terms like those in figure 23.2. We emphasise however that the Lagrangian so obtained is unique. Apart from the choice of a single coupling constant (which occurs in the diagrams of figures 23.1 and 23.2) there is no freedom. Indeed, something else rather nice happens. As we noted above the gluons will couple to each type of quark (u, d etc). At first sight we might think the coupling constants could be different: e.g. corresponding to figure 23.1(b) we could have terms $g\bar{\psi}_{u\mathrm{R}}\psi_{u\mathrm{G}}\mathrm{A}^{\mathrm{RG}}$ and $g'\bar{\psi}_{d\mathrm{R}}\psi_{d\mathrm{G}}\mathrm{A}^{\mathrm{RG}}$, etc, with g

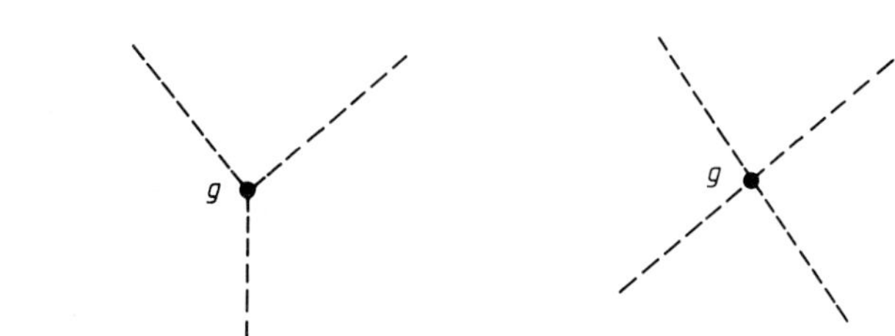

Figure 23.2 Showing the three-gluon and four-gluon interactions that are required when the colour-mixing symmetry is made local.

not equal to g'. (Here, ψ_{uG} is a green u quark, etc.) However, this is *not* possible because both couplings have to be equal to the 3 and 4 gluon coupling constants in figure 23.2. It is important to realise that the gluons themselves do not carry any labels referring to u, d, c etc; they just do not 'know about' such distractions. In particular they cannot change one quark type to another ($u \rightarrow d$ for example); thus they will be consistent with the observed conservation rules discussed in §22. Note that the change in *colour*, which can occur with gluon emission, does not have obvious experimental consequence, since observable particles are colourless.

We have developed a beautifully elegant theory, describing a lot of processes in terms of just one constant. Since it is a gauge theory, it is renormalisable. In the next section we shall justify the claim that this theory 'solves' strong interaction physics. Clearly such a theory deserves a name: by analogy with QED we call it 'quantum chromodynamics' or QCD.

§ 24 QCD and the real world

We now have a theory of strongly interacting particles containing just one free parameter. Is it correct? To answer this question we should calculate bound states, etc, and see if we can choose the parameter to obtain agreement with data. In particular we would like to be able to calculate the proton, the pion and all the other observed states. Regrettably this is not possible; the theory is much more complicated than QED and does not have a small parameter analogous to e^2/hc. There are however several features† that can be derived which give confidence in the validity of QCD. Most of these depend on a remarkable property of the QCD interaction, namely that it becomes effectively *stronger* as the interacting particles move apart or, which is essentially the same thing, becomes weaker as the energy of interaction becomes larger. The reason that these two statements are related is that short distances correspond (through the uncertainty principle, for example) to high momenta or high energy, and conversely. To understand why the interaction strength changes with energy we have to remember that we are speaking about a *renormalised* coupling constant (recall §20). The higher-order corrections have an (infinite) constant part together with a part that varies with energy.

This type of variation in the effective coupling constant is called 'asymptotic freedom', a name which reflects the fact that in the limit of infinite energy the particles become free. It has two experimentally important aspects. First, the increase of the coupling constant with distance hopefully explains the 'rule' introduced in the previous section that only colourless particles are ever seen. Quarks, and all other coloured, i.e. non-singlet, objects are strictly 'confined' within the observed (colourless) particles; attempts to separate them will always fail because of the increasing strength of the interaction. I say

† Apart from the rather persuasive one that it is so nice that it just *has* to be the answer!

hopefully because this confinement property has not been proved, although there are many indications that it is true. (Of course if free quarks *are* ever observed we will need to allow some exceptions to confinement.) One way of thinking of confinement is to imagine that the quarks inside a proton, for example, are joined by elastic bands that cannot snap. In some respects, a better way is to imagine them stuck in treacle so that they never come out cleanly; there are always $q\bar{q}$ pairs about so that the free quarks are not seen.

Second, the opposite extreme tells us that at short distances (probed by high energies) the quarks which are inside protons behave as though they are essentially 'free'. This justifies the many calculations that have been made on the assumption of free quarks. (It was, in the early days, very hard to understand why such calculations gave correct predictions when they completely ignored the strong forces needed to fasten the quarks together.) In recent years a lot of effort has been put into the calculation of 'QCD corrections' to the simplest approximation. The situation is somewhat confused but it does appear that, where believeable calculations are possible, these corrections are in agreement with what is seen. In particular, recent CERN results at the highest available energies show beautiful agreement with QCD predictions.

We now consider a 'theoretically' important aspect of asymptotic freedom. There is a crucial difference between protons as bound states of quarks and, for example, atoms as bound states of electrons and nuclei. In the latter, the mass of the bound state is approximately the sum of the constituent masses. In QCD however, the proton mass is much larger than the sum of the three quark masses (recall the figures given in table 22.1). Indeed it is believed that even if we let the u and d quark masses become zero then the proton mass would not change greatly (since M_{p} is about 100 times M_u and M_d we are not really far from this imaginary limit). At first sight this appears to cause a problem. In a Lagrangian without any mass terms, what can possibly determine the 'scale' that gives the mass of the proton, for example? (If we calculated and found the answer 3, say, we would not know if this were 3 tonnes, 3 kilograms or whatever.) Here we have another amusing property of asymptotic freedom. The only parameter of the theory is the dimensionless, renormalised coupling constant $g(E)$. (We have written it like this to emphasise that it varies with energy.) Now, since this becomes zero as E tends to infinity and is infinite when E becomes zero, there is no 'natural' place to define its magnitude.

What we can do, however, is introduce the energy at which it is equal to (say) unity

$$g(E_0)=1. \tag{24.1}$$

This energy of course defines a mass, through the relation $E=mc^2$, and this will be the mass scale of our theory. We expect the composite particles to have masses around this value. Note that these are what are called 'order of magnitude' arguments—a factor 10 either way is not important—so it would not matter if we replaced the 1 in equation (24.1) by, say, 1.7 or even 4.7. We simply need some indication of the energy at which the force changes from being weak to being strong enough to hold particles together.

Finally, we remember that our aim in this chapter is to explain the strong interactions between observed particles, i.e. the nuclear forces. Have we succeeded? We note first that all observed particles are 'colour singlets' (this is analogous to having zero charge in the QED case). Thus there is no direct gluon exchange between them. However, there will be residual forces, as already discussed for atoms at the end of §13. Here, they are much harder to calculate. The analogy with the van der Waals force between molecules is, in fact, somewhat misleading because gluons, unlike photons, are confined. This means that the residual interaction is sharply cut off. It has a range given by the inverse of the mass of the lightest particle that can be exchanged (cf remarks at the end of §19). Attempts at *real* calculations have been made—they are very approximate but appear to give the right answers.

We can summarise the situation by saying that QCD is elegant and simple in principle, has many of the right properties to fit observations, gives quantitatively correct results where reliable approximations exist, has not been confronted with a situation where it clearly fails, but is generally difficult and messy when attempts are made to produce accurate predictions and does not seem to be solvable by simple approximation methods. I believe it is correct, and expect evidence for this to accumulate slowly in the next few years. It is the explanation of strong interactions.

Summary of Chapter Two

The properties of atomic nuclei are easily understood when it is realised that they are made of protons and neutrons bound together by strong forces of short range (§§14, 17). The protons and neutrons are themselves members of a large class of objects which are all made out of quarks (§§17, 22). The forces which bind the quarks are obtained from a gauge field theory (§§18, 21) using the colour degree of freedom (§23). This theory, QCD, is a natural generalisation of QED, the quantum form (§§19, 21) of electromagnetism which was used in the first chapter. It explains nuclei, and, indeed, all strong interaction physics (§24).

This is a very satisfactory situation. We appear to have reached the end of our story because everything is explained. However, this is not the case; there exists a completely new interaction, much weaker than those so far discussed, which already shows itself in nuclei. We shall see in the next chapter that it leads us into new, exciting and mysterious aspects of physics.

Chapter Three

in which we meet some completely new effects,
with oddly different properties,
we are surprised to find many more particles,
we realise that the old ideas do not work,
so we use some clever new ones which appear to be verified.

§ 25 Weak interactions

We have explained in the previous chapter how some of the observed decays of nuclei (e.g. reactions (14.1) and (14.2)) can be understood as rearrangements of their constituent protons and neutrons. However, there are other decays that are not so simple. There are several examples where a nucleus with Z protons and N neutrons undergoes one of the reactions

$$[Z, N] \rightarrow [Z-1, N+1] + \mathrm{e}^+ + \nu_\mathrm{e} \tag{25.1}$$

or

$$[Z, N] \rightarrow [Z+1, N-1] + \mathrm{e}^- + \bar{\nu}_\mathrm{e}. \tag{25.2}$$

The particle ν_e which, together with its antiparticle $\bar{\nu}_\mathrm{e}$, is involved in these decays, is new to us. It is called the *electron-neutrino*, it has spin $\frac{1}{2}$ and very small (perhaps zero) mass. Because it is such an intriguing object we defer further discussion to the next section, which is devoted entirely to its properties.

In contrast to the previous decay modes of nuclei, those in (25.1) and (25.2) involve a change in the total numbers of protons and neutrons. Indeed, it is clear that they both occur through a neutron $\leftrightarrow$ proton transition, inside the nucleus, with emission of other particles. In particular, (25.1) is caused by

$$\mathrm{p} \rightarrow \mathrm{n} + \mathrm{e}^+ + \nu_\mathrm{e} \tag{25.3}$$

and (25.2) by

$$\mathrm{n} \rightarrow \mathrm{p} + \mathrm{e}^- + \bar{\nu}_\mathrm{e}. \tag{25.4}$$

The mass restrictions, arising from energy and momentum conservation as discussed in §16, will allow at most one of (25.3) or (25.4) to occur for free protons or neutrons. In fact the neutron mass is (just) sufficiently greater than the proton mass to permit its decay according to (25.4). This accounts for the fact that free neutrons are not normally found; they are unstable, with a lifetime of around 20 minutes.

For protons and neutrons inside nuclei it is the masses of nuclei which determine which, if any, will decay. In fact either (25.1) or (25.2) will occur whenever the mass of the original nucleus, $[Z, N]$, is greater than the sum of the masses of the final nucleus, $[Z \pm 1, N \pm 1]$, the electron and the neutrino. It is the existence of these decay modes, together with those discussed earlier (§14), which restrict the values of Z and N seen in naturally occurring nuclei.

From the above lifetime of the neutron, and knowledge of the masses involved, it is possible to calculate the strength of the basic interaction and conclude that it is many factors of ten weaker than either the strong interaction or the electric interaction—hence the name 'weak'. (Actually the extremely long lifetime of the neutron is due more to the close proximity of the neutron and proton masses than to the weakness of the interaction. Typical weak decays, which we shall meet later, tend to be associated with lifetimes around 10^{-8} s.)

Just as we explained the weak nuclear decays (25.1 and 25.2) in terms of decays of protons and neutrons (25.3 and 25.4), so in turn we expect these to be caused by decays of their constituents, e.g.

$$u \rightarrow d + e^{+} + \nu_{e}. \tag{25.5}$$

One possible way of obtaining this would be to put the appropriate interaction into the Lagrangian, which would then contain the term depicted in figure 25.1. Clearly this diagram, by suitable choice of the direction of the arrows (we remind ourselves that changing the direction of an arrow requires changing the particle to its anti-

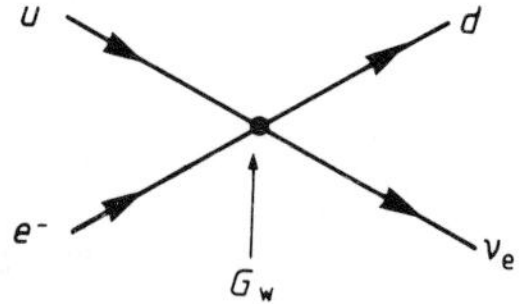

Figure 25.1 An interaction which would give the observed weak decays. Such an interaction, between four spin $\frac{1}{2}$ particles, is not allowed.

particle), gives rise to (25.5) and can also cause the proton $(u+u+d)$ or the neutron $(u+d+d)$ decays discussed earlier. The number G_w introduced in figure 25.1 is the appropriate (weak) coupling constant.

Although the interaction in figure 25.1 gives the observed results, it cannot be correct (though in many situations it is a very useful approximation). The reason is that such an interaction is not renormalisable. It is indeed worse than that; it would give rise to a cross section at high energy (*very* high) which would violate a rather basic requirement known as the unitarity limit, which is that the total probability of all things that can possibly happen must not exceed unity (an opinion poll that concluded that half the people would vote 'yes' and threequarters would vote 'no' would not be acceptable). It is just not permitted to have a point interaction involving four spin $\frac{1}{2}$ particles.

From what we have previously seen, we can readily guess a possible solution to this difficulty. The forces we already know about (QED and QCD) involve the interaction of spin 1 particles with spin $\frac{1}{2}$ particles, so it is natural to postulate the existence of a new spin 1 particle, which we call the W, and to introduce terms like those shown in figures 25.2(*a*) and (*b*) into the Lagrangian. Then the required interaction proceeds as in figure 25.2(*c*).

We must emphasise the difference between figures 25.1 and 25.2(*c*). In the former, the interaction occurs only when the particles are at the same point, whereas in figure 25.2(*c*) there is a non-zero range to the interaction. As discussed in §19 this range, R, is inversely proportional to the W mass $(R=h/Mc)$. At low energies, corresponding to wavelengths λ much greater than the range, the two interactions cannot be distinguished. However, as λ becomes of the same magnitude as R, the cross section arising from figure 25.2(*c*) will fall below that arising from the point interaction, and will thereby remain consistent with the unitarity limit noted above. In practice, because the W mass is very large (see below), this effect of the finite range has not yet been observed experimentally. On the other hand the necessary measurements to verify that W has spin 1 have been made.

Early in 1983 W particles were observed for the first time. They were created through $\bar{\text{p}}$p collisions at CERN. This experimental confirmation of their existence and, even more spectacularly, the fact that their observed mass agrees with that which had been predicted beforehand, are impressive confirmations of the essential truth in the ideas we are studying.

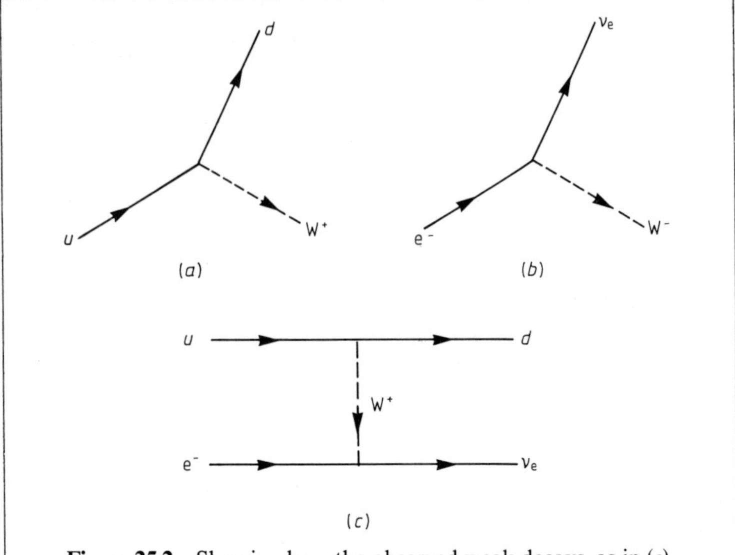

Figure 25.2 Showing how the observed weak decays, as in (*c*), are made from basic terms (*a*) and (*b*) in the Lagrangian.

Before we discuss the theoretical basis for this prediction of the W mass, we must look at some other weak interaction effects. Here we have to introduce another new particle into our story: the μ (called the 'mu' or 'muon'). This is an exact replica of the electron, except that it has a mass about 200 times greater ($M_\mu \sim 0.1 M_p$). The late introduction of the muon into our story is a serious violation of historical order, since it was the first particle, beyond the basic 'building blocks' (e, p, n), to be found. It was, then, a mystery: why should nature wish to duplicate the electron? Who needs the muon? It is still a mystery. We do not have any answers to these questions. Indeed, as we shall see, the mystery has become greater.

The muon interacts with the W in the same way as the electron does, but with its own neutrino. Thus we have a term like figure 25.3(*a*) in the Lagrangian. Combining this with figure 25.2(*b*) we obtain the W exchange process depicted in figure 25.3(*b*). Because the μ mass is much greater than the electron mass (and the ν_μ is again very light), this process occurs as the decay of the muon

$$\mu^- \rightarrow e^- + \bar{\nu}_e + \nu_\mu \tag{25.5}$$

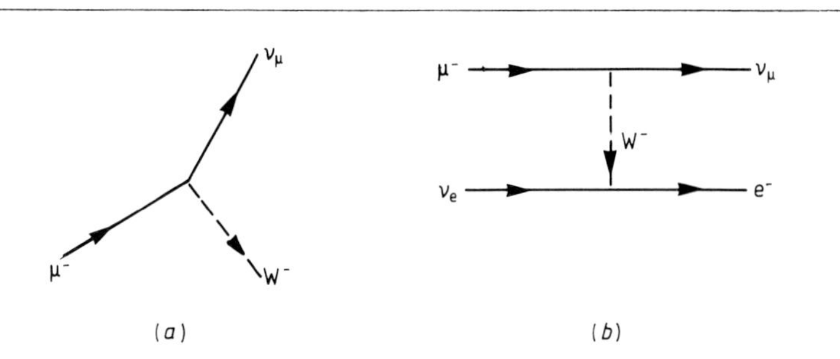

Figure 25.3 In (*a*) we show the basic interaction of the muon and its neutrino with the W. In (*b*) we see how this contributes to μ decay (equation (25.5)).

with a lifetime of about 10^{-6} s. To obtain this lifetime, the coupling constant in figure 25.3(*a*) has to be the same as those in figures 25.2(*a*) and (*b*) except for a small modification, to which we shall come in due course (§27).

There is a lot more to weak interactions, but the time has come to have our promised section on the neutrinos.

§ 26 The neutrinos

These are undoubtedly among the real stars of particle theory. Because they do not participate in strong or electromagnetic interactions they are, as far as we know, the most 'free' objects in the Universe. The only interactions which affect them (e.g. those in figure 25.2 etc) are so weak that neutrinos can travel through thousands of miles of matter with only a small chance of a collision. Indeed, neutrinos formed near the centre of the Sun can easily escape. Some come to the Earth and by detecting them (not easy!) we can 'see' into the middle of the Sun. Of course, most of the Sun's neutrinos that reach the Earth pass straight through.

The history of our understanding of the neutrino is worth mentioning. In an event such as reaction (25.4), only the proton and electron can be observed by detectors. Thus, the *apparent* event is $n \rightarrow p + e^-$. In such a process the energy of the electron, for example, would be uniquely fixed from the masses and the laws of conservation of energy and momentum. However, when this process was observed, the electron energy was seen to vary from event to event. To explain this violation of accepted conservation laws the unseen neutrino was postulated; its only role was to carry away some unwanted energy and momentum. There was much debate as to whether the particle was 'real' or simply a construction of theoretical physicists. Certainly there was general agreement that it would never be directly observed.

Detailed measurements of the energy and momentum in processes like (25.4) have gradually improved the limits on the neutrino mass. Most are compatible with its being zero, and a limit

$$m(\nu_e) < 10^{-4}\, m_e \qquad (26.1)$$

is now established, where m_e is the electron mass. For the ν_μ such accurate results are not possible; all we know is that it is lighter than the electron.

Neutrinos have now been directly observed (contrary to the

previously noted 'agreement'), and experiments involving neutrino beams are part of the contemporary scene in particle physics. Even though the cross section for a collision is small, an intense enough beam, a large enough target and a long enough wait produce neutrino events. One of the most important results to come from the first neutrino experiments was the fact that, as we have indicated by giving them a different name, the ν_e and ν_μ are *different*. This was deduced because it was known that the neutrinos in the beam used in the experiment were mainly produced in association with muons; in fact they came from the decay:

$$K^+ \rightarrow \mu^+ + \nu_\mu \tag{26.2}$$

(see §27 for further discussion of this). When these interacted with matter (i.e. protons and neutrons), the observed events were

$$\nu_\mu + n \rightarrow p + \mu^- \tag{26.3}$$

and *not*

$$\nu_\mu + n \rightarrow p + e^- \tag{26.4}$$

whereas, if the ν_μ and ν_e are the same, these two processes should occur with approximately equal probability. Thus, we conclude that the ν_μ and ν_e are different particles.

So far in this chapter we have ignored one really new feature of weak interactions—they violate parity. We can best explain what this means by considering the way the neutrino enters the weak interaction. First, we recall §12 and remark that a neutrino, being a spin $\frac{1}{2}$ particle, can have a spin $\pm\frac{1}{2}$ in any chosen direction. For this direction we shall choose the velocity of the neutrino. Notice that for most particles this would not, perhaps, be a good choice, because there would be the possibility of the particle being at rest. However, for a massless neutrino (pretend $m_\nu = 0$ for the present) this possibility is not allowed, since it must always move with the velocity of light†. Thus, the spin is either along the direction of velocity or against the direction of velocity. We refer to these as left-hand or right-hand neutrinos respectively (see figure 26.1). Notice again that a left-hand

† The fact that the massless photon always has this velocity was of course the origin of the theory of special relativity (§15). That the same thing is true for other massless particles can be seen directly by comparing equations (16.1), (16.2) and (16.3) (with $m = 0$).

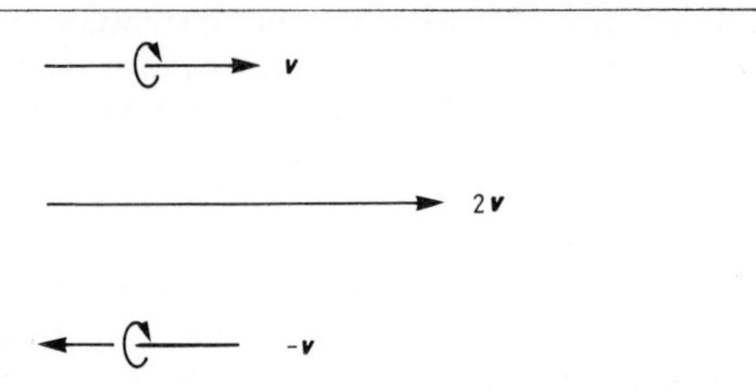

Figure 26.1 In the top picture we see a left-hand particle moving with velocity $\boldsymbol{v}$, as seen by observer 1. The second observer has velocity $2\boldsymbol{v}$ relative to the first, as shown in the second picture. Thus the particle appears to observer 2 as shown in the bottom picture; its velocity is $-\boldsymbol{v}$, its spin direction is unaltered, and it therefore appears right-handed. This cannot happen for a massless particle.

neutrino, for example, is seen as a left-hand neutrino by *all* observers regardless of their velocity. It is not possible to have an observer moving *faster* than the neutrino (nothing can move with a velocity faster than c) so the direction of the neutrino's velocity cannot be reversed. This would not be the case for particles with mass, as we illustrate in figure 26.1. Of course, the observers have to agree on the definition of their 'left hand', i.e. they all need the same model or convention†. This is the point mentioned already below equation (10.2).

Now comes the new feature of weak interactions. *Only left-hand neutrinos participate in them.* Note that this means that right-hand neutrinos never interact, which, if true, is equivalent to saying that they do not exist. Because of the need of a model to state this rule, or, which is the same thing, because an observer who views the world in a mirror would see a different law, i.e. only right-hand neutrinos would exist, this property of weak interactions is described as 'parity violation'.

What we are doing here can be expressed in terms of the Lagrangian in the following way. We denote the wave function

† It is usual to point the thumb in the direction of the velocity; then the fingers give the direction of the spin. Check this for the particles shown in figure 26.1.

associated with the neutrino by the symbol ν and we separate this into two functions ν_L and ν_R which we can associate with the left-hand and right-hand neutrinos. To explain precisely how this is done would take us too far into the mathematical details, so we merely state the important results which are that the kinetic energy term in the Lagrangian takes the form '$\overline{\nu_L}\nu_L + \overline{\nu_R}\nu_R$', whereas the mass term becomes '$m(\overline{\nu_L}\nu_R + \overline{\nu_R}\nu_L)$'. Thus the presence of a mass term requires the existence of both L and R neutrinos, in agreement with the fact noted above that only for a zero mass neutrino does it make sense to postulate that there are no right-hand neutrinos. (Notice that the bar in $\overline{\nu_L}$ covers *everything*; it defines the antiparticle of the left-hand neutrino so, in accordance with the usual rule for antiparticles—that we change everything except the mass—it gives a right-hand object.)

Even when mass terms are present it is, of course, still possible to define the left-hand and right-hand states, and to use only the left-hand states in the weak interaction. This, which corresponds to maximal violation of parity, is what nature appears to do—not only for the neutrino, but also for the quarks, the electron, the muon, etc. This 'etc' will form the topic of our next section.

§ 27 Families

As we have seen, the W couples to pairs of spin $\frac{1}{2}$ particles (e.g. figure 25.2(*a*) and (*b*), figure 25.3(*a*)). The complete list of such pairs, as far as they are known in early 1985, is shown in table 27.1, where the numbers in parentheses are the particle masses in units such that the proton mass is 1.

The leptons do not carry the colour label and therefore do not see any strong interactions. The quarks all come in three 'colours', i.e. they are triplets under the QCD transformations. As we have seen, this means that they are always confined inside other, colour-singlet, particles, e.g. protons, pions, etc.

Each horizontal foursome in the table is called a 'family', and the table contains three separate families. Apart from the masses, the three families are identical. *We have no explanation of this replication; it is the modern form of the muon puzzle mentioned at the end of §25 and is certainly one of the major puzzles still awaiting a solution.*

Here we must digress to discuss the 'top' quark, which is included in the table for rather obvious reasons, although its existence has only recently been given any experimental support. The place where it was once thought most likely to be seen is in the third experiment

Table 27.1 The three known families of quarks and leptons.

	Leptons		Quarks	
Charge	-1	0	$\frac{2}{3}$	$-\frac{1}{3}$
First family	electron e^- (0.0005)	e neutrino ν_e (0 ?)	up u (0.005)	down d' (0.01)
Second family	muon μ^- (0.1)	μ neutrino ν_μ (0 ?)	charmed c (1.5)	strange s' (0.15)
Third family	tau τ^- (1.8)	τ neutrino ν_τ (0 ?)	top t (about 40)	bottom b' (4.7)

mentioned in §1, where it could be produced through the collision of electrons on electrons,

$$e^- + e^+ \rightarrow t + \bar{t} \tag{27.1}$$

as shown in figure 27.1.

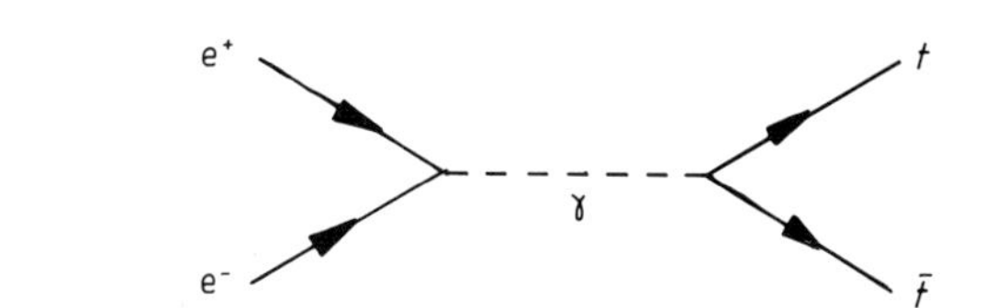

Figure 27.1 Lowest order diagram for the production of a $t+\bar{t}$ pair by an e^+e^- collision.

Actually, since free quarks cannot exist, the $t+\bar{t}$ must recombine to form groups of other strongly interacting particles (called 'jets'). The quantity that is measured is the ratio of the probability for producing such strongly interacting particles compared to the probability of producing a $(\mu^+ + \mu^-)$ pair. Since the latter are produced through the same diagram as that in figure 27.1, differing only in the charges at the right-hand vertex, this ratio is easy to calculate. A given quark contributes Q^2 to this ratio, where Q is the charge on the quark (recall that the muon has charge -1). Thus if we include only the u, d, c, s and b quarks we find a ratio

$$3[(\tfrac{2}{3})^2 + (\tfrac{1}{3})^2 + (\tfrac{2}{3})^2 + (\tfrac{1}{3})^2 + (\tfrac{1}{3})^2] = \tfrac{11}{3}$$

where the factor 3 comes from the 3 colours of each quark. This is the value observed experimentally, and it implies that *no* top quark contribution is present, i.e. that $t+\bar{t}$ production is forbidden by energy conservation:

$$E < 2M_t \tag{27.2}$$

where E is the available energy. Current experiments have E up to 43 M_p, so we conclude that there are no more quarks with mass below half this value. It is easy to see that, whenever E becomes greater than $2M_t$, the ratio should rise to 5 (hence the remark regarding this experiment in §1).

The recent positive evidence for the top quark, and for the quoted value of its mass, has come from the 'SPS' machine at CERN which collides protons on antiprotons. This machine has also been responsible for the experimental observation of the W and Z^0 of §25. Several events which appear to be examples of the process shown in figure 27.2 have been seen. Here the W decays to a t plus a $\bar{b}$. The t then

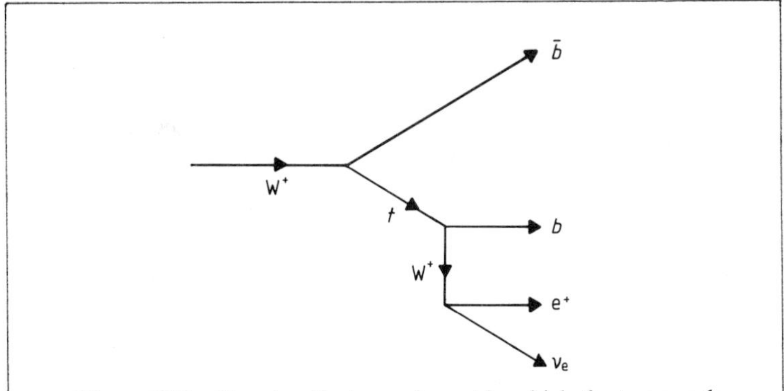

Figure 27.2 Showing the type of event in which the top quark (t) has recently been discovered at the CERN Collider.

subsequently decays, by a reaction analogous to that of (25.5), into a b quark plus a neutrino and a positron. It is this positron, together with the 'jets' formed by the b and $\bar{b}$ quarks that we actually observe, so some work is necessary in order that we can be sure that the events are being interpreted correctly. Confirmation of this discovery is expected to come from the machine run with improved detectors taking place in late 1984.

We now return to table 27.1. The observant reader will be wondering why we have used d', s' and b' in the last column of the table rather than just d, s and b. The reason is another fascinating aspect of weak interactions. The states that participate in the pairs with the u, c and t are not exactly the previously defined quarks but are particular combinations of all three. Thus, for example,

$$d' = xd + ys + zb \tag{27.3}$$

where x, y and z are numbers such that

$$x^2+y^2+z^2=1. \tag{27.4}$$

What does this mean? Consider the coupling of the W to the pair (u, d'). Using equation (27.3), we see that this yields three different processes, as shown in figure 27.3. The first of these we have already

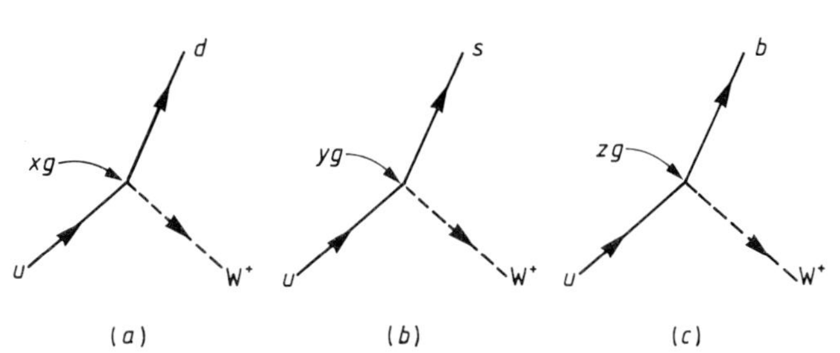

Figure 27.3 Showing how the $ud'W^+$ interaction yields three separate terms when equation (27.3) is used for the d'. The diagram (a) shows the reduced coupling to ud, diagrams (b) and (c) show violation of strangeness and bottom conservation.

met (figure 25.2(a)). Note, however, the presence of the extra factor x which reduces the coupling to (u, d) (this is the small modification referred to already towards the end of §25). In fact, x turns out to be close to unity, implying, from equation (27.4), that y and z are quite small.

The second term in figure 27.3 is new. It involves a change from one family to another. In fact, it is this diagram which is responsible for the small violation of strangeness conservation mentioned in §17. For example, we show in figure 27.4 a diagram which causes the decay

$$\Sigma^+ \rightarrow p+\pi^0 \tag{27.5}$$

thereby giving the Σ a lifetime of about 10^{-10} s.

Another interesting example is K decay:

$$K^+ \rightarrow \mu^+ + \nu_\mu \tag{27.6}$$

and

$$K^+ \rightarrow e^+ + \nu_e \tag{27.7}$$

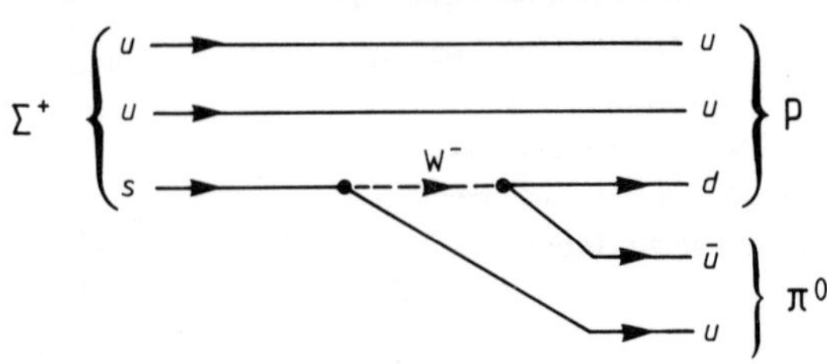

Figure 27.4 A decay which violates conservation of strangeness. The Σ^+ with strangeness 1 decays to a proton and a π^0, both with zero strangeness.

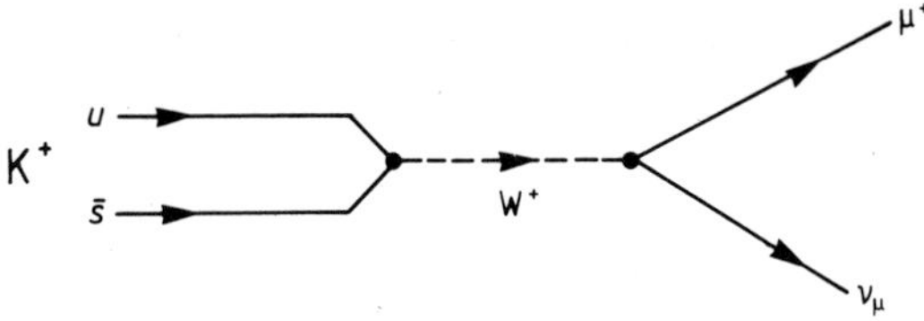

Figure 27.5 The strangeness-violating decay of the K^+ into a $\mu^+\nu_\mu$ pair.

which takes place through diagrams like that in figure 27.5. It is amusing to see why the decay into a muon (equation (27.6)) is in fact much more likely than into an electron (equation (27.7)), although the basic couplings are identical. From the point of view of an observer to whom the K is at rest, the other two particles will move in opposite directions (conservation of momentum) as depicted in figure 27.6. On the other hand, as we saw in the previous section, the ν will be left-hand and the $\mu^+(e^+)$ would be right-hand *if it were massless.* Now a left-hand particle and a right-hand particle moving in *opposite*

Figure 27.6 Showing why a spin zero particle cannot decay into one left-hand and one right-hand particle.

directions are spinning in the *same* direction (see figure 27.6). Thus their spins add to give total spin 1. Since the K has spin zero the process is forbidden by conservation of spin. Thus the K decay can only occur because the μ and e are not massless. For the e the zero mass approximation is quite good (the appropriate parameter is m_e/m_K, which is about 0.001) and the decay is strongly suppressed, whereas for the μ (with $m_\mu/m_K = 0.2$) the spin suppression is much less. Hence the K decays mainly into muons.

We now know a lot of facts about weak interactions, so we turn to the problem of trying to explain them. This will be our aim for the remainder of this chapter.

§28 Composite quarks and leptons

This section is in the correct place for the logical continuation of our story. However, from the historical point of view, it should come much later. It is also, in contrast to everything else in the first three chapters, very speculative, and for this reason belongs more naturally to the fourth chapter. Readers may choose to postpone it until later, since its content will not be required in subsequent sections.

The weak interaction is a short-range force, i.e. the particle responsible (the W in figure 25.2(c) for example) is massive. This is reminiscent of the situation we encountered with nuclear forces, so it is natural to consider the possibility of a similar solution. We recall that in chapter 2 we found that the observed interaction was not 'fundamental' but was, rather, the residual effect of a simple gauge theory interaction between confined constituents. To solve the weak interaction situation in the same way, we must therefore postulate that the quarks and leptons are composites of other, yet more basic, objects.

Before we attempt this we note that it could well be very rewarding. We might expect, by analogy with QCD, to have to introduce one new constant (the coupling or mass scale). With this one constant we should, in principle, be able to derive the masses of all the bound states and indeed reproduce table 27.1; it should also yield the weak interaction—and hence the mass of the W. We would then appear to be very near the end of our story. There is however (at least) *one very big obstacle* to such a solution.

To understand this obstacle we must consider the 'size' of the quarks and leptons. This can be measured by considering their interaction with photons. For example, high energy photons produce pairs of electrons:

$$\gamma \rightarrow e^+ + e^- \tag{28.1}$$

as shown in figure 28.1. This, of course, is the basic interaction of

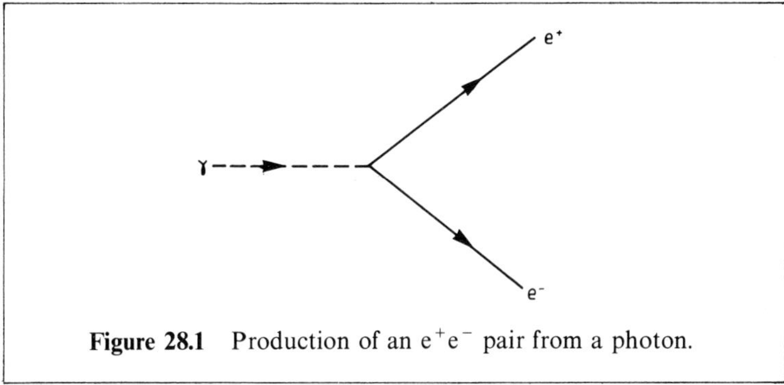

Figure 28.1 Production of an e^+e^- pair from a photon.

electromagnetism and it can be calculated. Normally, in making the calculation, we *assume* that the electron is a *point* charge, i.e. has zero radius. It is possible to correct for the effect of a non-zero radius. A factor is introduced which is approximately 1 at low photon energy, where the size has little effect, but which decreases as the energy increases. (This agrees with the remark in §14 that we need a sufficiently high energy to 'see' small structures.) So far, experiments have shown no evidence for any deviation from a point electron, thereby yielding an upper limit to the radius. In fact, the best limits on the electron, and muon, radii come from related, but somewhat more involved, considerations. These show us that the radii are less than about 10^{-20} m. For quarks, the upper limits are similar, though perhaps a little larger.

Now, why do such small radii cause problems? To understand this we need to remember the uncertainty principle. If the constituents of leptons are confined within a radius less than R_0, say, then they will have a momentum uncertainty greater than h/R_0. Since the average momentum must be at least as large as the uncertainty (reasonably obvious, but this can be made precise), we find that the kinetic energy of the constituents is at least 10^4. (As usual we use the proton mass as our unit of mass and its rest energy, M_pc^2, as our unit of energy.) The rest energies of the leptons and quarks are very much smaller (e.g. 0.5×10^{-3} for the electron), so this kinetic energy of the constituents must be cancelled by 'potential energy' (due to the forces) to an accuracy of, for the electron, about 1 part in 10^7. This does not seem believeable!

We can summarise the problem in the following way: the 'scale' of the radius of the bound state and of the mass are all determined by the same quantity (the mass scale of the binding interaction) so we would expect:

$$\text{`Typical mass'} = \frac{h}{c} \times \frac{1}{\text{`Typical radius'}} \tag{27.2}$$

whereas, in fact, we find masses at least 10^{-6} smaller than this.

It is unreasonable to expect that this sort of thing should occur 'accidentally', especially as it happens for several particles, so we seek a reason. In fact, such a reason is not hard to find and is related to the point we noted before about the neutrino: if there is no right-hand neutrino then the neutrino mass has to be zero. Let us suppose that the constituents of quarks and leptons are *massless*; then, as we saw in §26, the quadratic term in the Lagrangian can be written '$\overline{\Psi_L}\Psi_L + \overline{\Psi_R}\Psi_R$'. This has a global symmetry which is *new*: we can alter the phase of the L and R parts of Ψ separately. Previously we have always treated them alike. More generally, when we have several Ψ fields, we can mix the L and R particles independently, without affecting the Lagrangian. We refer to such symmetries as 'chiral' symmetries. Mass terms, of course, clearly break such chiral symmetries (e.g. if we alter the phase of Ψ_L, but not of Ψ_R, then we change the mass term $m\overline{\Psi_L}\Psi_R$). Thus, a theory with chiral symmetry has no mass terms. This applies not only to the original Lagrangian and its fundamental fields, but also to all bound states. (There is, in fact, a way of obtaining mass terms when two different states have the same mass—we can, however, ignore such peculiarities.) Thus, with our assumption of zero-mass constituents, and provided we do not add interactions that violate the symmetry, we obtain exactly massless composite electrons, etc. This looks as though it might solve our problem. Compared to the scale (h/R_0c) the observed masses are nearly zero. We would, of course, need later to think about the origin of these small masses. Unfortunately, however, something is wrong with all this.

Chiral symmetries, such as we have just discussed, are generally mathematically inconsistent. The reason for this is that one can calculate some particular quantities in terms of the original set of massless particles, or in terms of the bound states. They have to be equal and in most cases they are not. In practice this presumably means that the symmetry, even if it is there in the Lagrangian, must 'break'. Mechanisms through which this might happen will be discussed in §30. For the present, we note that we already have an example where such symmetry breaking must occur. Recall that in QCD we saw that the proton, for example, has a mass even if its constituents (u, d quarks) are massless so that the Lagrangian has a chiral symmetry. This can only happen if the symmetry is broken.

A set of conditions which have to be satisfied in order that a chiral symmetry remains unbroken has recently been given by 'tHooft. No realistic theories which satisfy these conditions exist.

Sometimes in the history of physics the best way round a problem has been to ignore it—hoping that its solution will come later. In this spirit many composite models have been suggested. We discuss now one of the simplest and most elegant: the 'rishon' model. The rishons are two spin $\frac{1}{2}$ particles, T and V, with charges $\frac{1}{3}$ and 0 respectively. Then we can play the quark model game again and construct a family of three-rishon states, as shown in table 28.1.

Table 28.1 The rishon content of the first family. (This table looks, to me, so simple that I sometimes feel that it *has* to be correct).

	Charge	Colour
$e^+ = \mathrm{TTT}$	1	singlet
$u = \mathrm{TTV}$	$\frac{2}{3}$	triplet
$\bar{d} = \mathrm{TVV}$	$\frac{1}{3}$	triplet
$\nu_e = \mathrm{VVV}$	0	singlet

The charges obviously come out correctly. In order to bind the rishons we naturally give them a *new* label which can take three values and make the appropriate symmetry into a local gauge symmetry. This is completely analogous to colour and QCD—so we call it hypercolour and QHCD. The quarks and leptons are of course (unique) hypercolour singlets. To obtain colour (the 'old' colour) for the quarks, the rishons must also carry colour labels. It turns out (and this requires a little knowledge of group theory) that both T and $\bar{\mathrm{V}}$ have to be colour triplets—then the leptons can be singlets and the quarks triplets as required.

This model has many nice properties, but it does illustrate a further problem (which may not really be a problem) of composite models.

Quarks and leptons are made of the same constituents, so they can change into each other; in particular, baryon number (§17) is not conserved. The most dramatic illustration of this is the possible decay of the proton

$$p \rightarrow \pi^0 + e^+ \tag{28.3}$$

which can proceed as shown in figure 28.2. Note that the existence of such a decay would imply that *matter* is unstable! If the process exists, we are fortunate that at least it is rare.

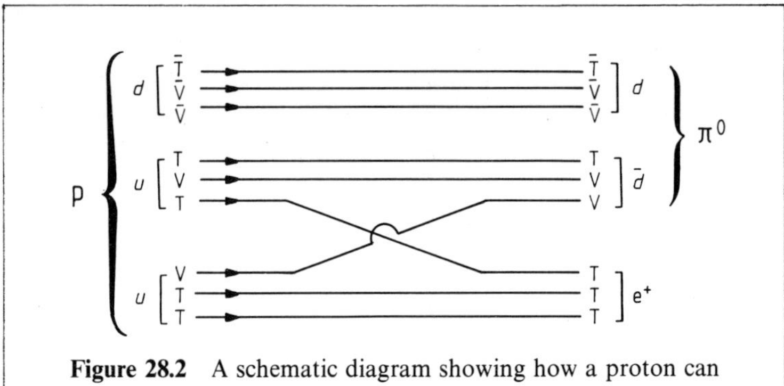

Figure 28.2 A schematic diagram showing how a proton can decay in the rishon model.

It is even possible to estimate the lifetime for the proton from this process. The value obtained depends on the details of the model and is very dependent on the radius of the quarks. In many models, consistency with the known proton lifetime (greater than about 10^{32} years, as explained in §31) requires the radius to be many factors of ten smaller than the limits previously noted. This of course only emphasises the 'scale' problem. It is certain that the apparent absence of proton decay is a severe constraint on composite models.

Finally we should ask how the weak interactions arise in composite models. Presumably they should be understood as 'residual effects' of the hypercolour interactions, in the same way that nuclear forces are a residue of the colour force of QCD. The observed reactions would occur through a rearrangement of rishons, e.g. figure 25.1 would proceed as in figure 28.3. However, it is necessary to ask whether such a model would correctly predict the established properties of the

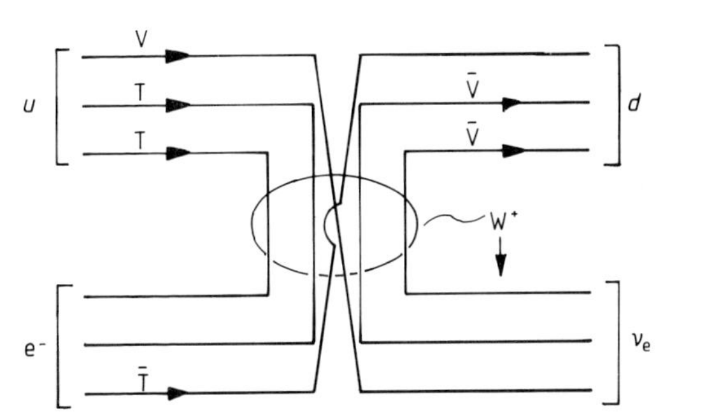

Figure 28.3 Showing how weak interactions can arise through the exchange of rishons. By comparison with figure 25.2(*c*), we see that the W^+ is a (TTTVVV) composite†.

weak interactions. At the end of the next section we shall see that there are both 'yes' and 'no' indications.

To summarise this section we can say that it is possible (even 'likely') that quarks and leptons are composite, but this compositeness is well hidden and there are at present no models which allow reliable predictions of their properties. The only experimental support for compositeness is the *very* tentative indication mentioned in the footnote on page 138.

† We remind readers here that reversing an arrow changes particle to antiparticle; so the electron which appears here as $\bar{T}$ together with two T particles with opposite arrows is really $3\bar{T}$, as required.

§ 29 The Salam–Weinberg model

As we have already remarked, the previous section is not in historical order. Indeed, even before the success of QCD was realised, a model based on local gauge invariance had been proposed for weak interactions. This predicted effects which were unknown at the time, but which have since been verified. The theory, due to Glashow, Salam and Weinberg, was imaginative and has been proved to be one of the great predictive theories of physics.

It is clear that figure 25.2(*c*), which is the origin of the weak interaction, is formally similar to QED and QCD; so, for the moment, we ignore the fact that the W, introduced in §25, has a mass, and we try to explain it as the spin 1 particle associated with a local gauge theory, just like the photon of §20 and the gluons of §23. We immediately meet a new feature. Whereas the gluon couples to pairs of the same electric charge (figure 23.1, for example) so that it does not itself have any charge, the W couples to particles of different charge (e.g. figure 25.2) and, because of charge conservation, must therefore be a charged particle ($W^{\pm}$ have charges $+1$ and -1, respectively) which couples to the photon. Hence, unlike the situation with QCD and QED which are completely independent gauge theories, it is not possible to consider the gauge theory of weak interactions without at the same time incorporating electromagnetism. In other words, to succeed in our aim, we must 'unify' weak and electromagnetic interactions. This looks as though it is real progress, so, eagerly, we proceed.

We wish to consider mixing of *two* objects, i.e. the pairs in table 27.1. To be specific we shall consider mixing the electron with its associated neutrino. The appropriate mass term in $\mathscr{L}$ has the form $m_e\bar{\psi}_e\psi_e + m_\nu\bar{\psi}_\nu\psi_\nu$, which is invariant under this mixing only when $m_e = m_\nu$. Thus we shall work in the approximation in which both these masses are zero.

Any global (i.e. constant) transformation among these two fields can be constructed from 4 basic ones, combined in suitable proportions: (i) $\psi_e \to \psi_\nu$ and $\psi_\nu \to \psi_e$, (ii) $\psi_e \to \psi_\nu$ and $\psi_\nu \to -\psi_e$, (iii) a phase change equal but opposite in sign for ψ_e and ψ_ν and (iv) a phase change equal for ψ_e and ψ_ν. (Mathematicians would express this by saying that the transformations being considered here, called U(2), have four generators.)

Now comes the key point. It is possible to make a local symmetry out of the first three of these, with one coupling constant (g_1), and out of the last one, with, in general, a different coupling constant (g_2). Thus, if we separate in this way, we essentially have *two* independent gauge theories. So, if we associate the second of these, i.e. the equal phase change, with the photon, then electromagnetism will be separate. However, the equality of the phase change implies that the *electric charges are identical.* This was correct in the QCD case†, but is not what is required here. Rather, we want to associate the photon with a suitable sum of (iii) and (iv) so that we have a phase change for the electron, but none for the neutrino (since it has *zero* charge). Thus the photon coupling (the electron charge) will be a combination of g_1 and g_2. There will also be another spin 1 particle, associated with another combination of (iii) and (iv), coupling to the electron and neutrino and not changing them into each other; this is called the Z^0. We try to explain all this is figure 29.1. In (*a*)–(*d*) we show the four couplings noted above, where we have named the spin 1 particles the $W^\pm$, W^0 and B^0 respectively. In (*e*) and (*f*) we have replaced the W^0, B^0 couplings by couplings to the photon and the Z^0, both of which are particular combinations of W^0 and B^0. Notice that the two sets (W^0, B^0) and (photon, Z^0) are completely equivalent, so we have not actually done anything at this stage.

Figure 29.1 also shows something else that will be necessary if we are to find a satisfactory theory. As we have noted, the $W^\pm$ only couple to left-hand states. Thus, the first three of our transformations only operate on the left-hand part of the fields. The B^0, on the other hand, couples to both. We must of course make sure that the photon couples equally to left- and right-hand electrons (since electromagnetism conserves parity).

To proceed we must somehow give non-zero masses to the $W^\pm$ and to Z^0; this will then uniquely separate the Z^0 from the photon, and the description in figures 29.1(*c*) and (*d*) will no longer be the same as that

† Different colours have the *same* charge.

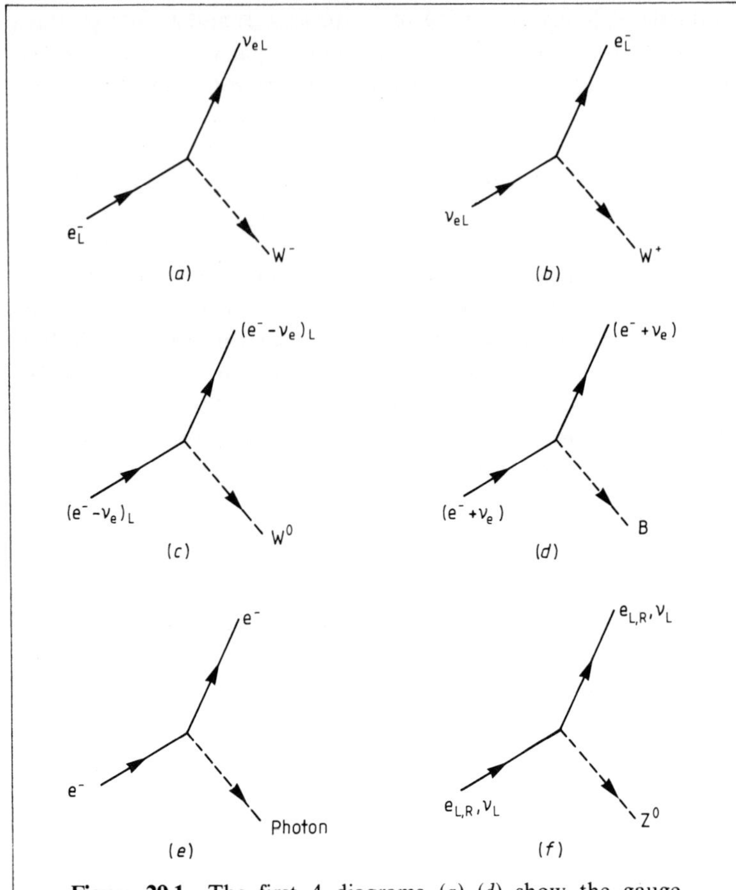

Figure 29.1 The first 4 diagrams (*a*)–(*d*) show the gauge interactions associated with mixing of e and ν_e. In the last two we have replaced (*c*) and (*d*) by equivalent interactions with the photon and the Z^0, which are suitable combinations of W^0 and B^0.

of figures 29.1(*e*) and (*f*). However, we have previously stated, correctly, that such masses are incompatible with our local gauge invariance. The only way out of this impasse is to abandon the invariance, and, remarkably, it appears as though there is a way of doing this which preserves many of the above results and which, in particular, still gives a renormalisable Lagrangian. The actual Lagrangian is very complicated and, as we shall see, contains some new fields. However, it is still sufficiently restrictive to allow some definite predictions. We postpone a more detailed look at the Lagrangian to the next section, where we shall see that it is obtained by a new, remarkable, procedure, which might play a vital role in physics.

The most obvious new feature which arises from our attempt to make the weak interaction into a gauge theory is the existence of the Z^0, interacting with a 'neutral current' in distinction to the charged currents with which the $W^{\pm}$ interact. This prediction of an additional weak interaction is an almost inescapable consequence of the gauge theory idea. That such interactions had not been observed when this fact was realised was not a discouragement, because they are not easy to observe. Whereas the $W^{\pm}$ give rise to effects which cannot be obtained by any other means, so even if they are small they can be detected, the Z^0 does not cause any change of particle type and its effect is normally overwhelmed by strong or electromagnetic interactions. The only exception to this occurs with neutrinos, which only take part in weak interactions. So it was with neutrino beams that these neutral current interactions were first seen; the process shown in figure 29.2 being observed at CERN in 1973.

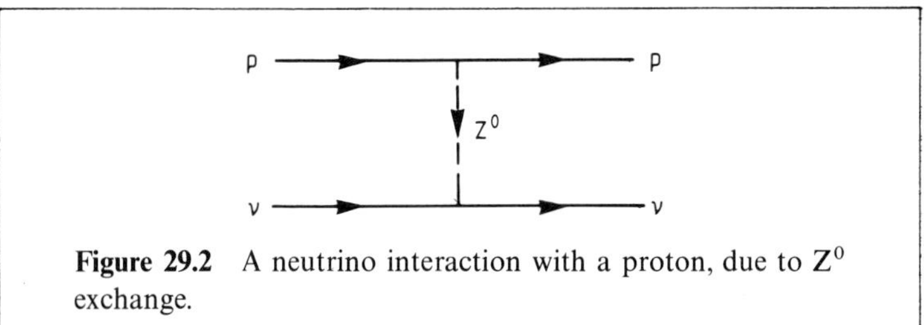

Figure 29.2 A neutrino interaction with a proton, due to Z^0 exchange.

The theory does not predict the magnitude of the Z^0 contribution, but once this had been established in one process, it became possible to predict the way the Z^0 couples to both L and R states. These predictions have been verified. Further, the theory *predicts* the masses of the $W^{\pm}$ and Z^0 particles as about 80 M_p and 90 M_p respectively.

It is only very recently that such energies have become available and a great deal of effort has gone into the search for the W and Z at the CERN SPS (which collides protons on antiprotons). Early in January 1983, several events which seem to imply the presence of Ws (seen through their decay into an electron plus a neutrino) were announced. During May of the same year the decay products of the $Z \rightarrow \mu^+ \mu^-$ and $e^+ e^-$ were seen. In both cases the measured masses agree with the above prediction†. We are clearly moving in the right direction!

Before we look a little further at the theoretical foundation of the Salam–Weinberg model, we make an important remark. The success of the predictions does not necessarily mean that the details of the model are correct. There is an interesting reason for this. Provided we do not introduce many more new, light, particles, and provided we insist on renormalisability, the Salam–Weinberg Lagrangian is probably unique. Thus the predictions have to be correct even if our particular derivation is wrong. Suppose, for example, that the real world *has* many more particles—as might well be the case if the quarks and leptons are composite (see §28). Since these particles have not been seen they must be very heavy (mass greater than about 100 M_p, for example). Similarly, we know that effects of compositeness certainly do not show up till this sort of energy or higher. In this situation we should be able to write down an *effective Lagrangian*, containing only the particles we know about and having no effects due to their composite structure, which will give correct predictions provided we keep to 'low' energy experiments. Now comes the important point: this effective Lagrangian itself has to be renormalisable (this result is 'plausible'—there are also some relevant theorems). Thus, the effective Lagrangian has to be that of the Salam–Weinberg model.

This of course is the favourable part of the answer to the question posed at the end of the previous section. Provided the observed particles are the low-mass composite states, and provided the

† The production rates of the Z^0 and $W^{\pm}$ are also approximately correct. In this connection it is worth noting that some of the Z^0 are 'lost' because they decay into $\nu + \bar{\nu}$ which are not seen. This argument can be used to show that there cannot be too many *light* neutrino types. An upper limit of about 20 is obtained. Thus, if all neutrinos are massless, there cannot be more than about 20 families. As we shall see later, cosmological arguments in fact give a much smaller limit.

composite model actually has $W^{\pm}$, Z^0 spin 1 composite states which are also in the low-mass sector, all will be well. The second proviso here, however, is serious. Although we might be able to use chiral symmetry (perhaps partially broken) to keep spin $\frac{1}{2}$ objects in the low-mass region, there is no corresponding symmetry to prevent the spin 1 particles from becoming heavy, i.e. having masses on the scale of the inverse radius (equation (28.2)). Of course it might *just* be possible to regard 80 and 90 proton masses as being on this scale, but it seems doubtful. Even if we did, of course, there would be a problem as to why such objects should appear at all in the 'low-energy' effective Lagrangian.

This brings us to the end of a rather large and difficult section, but we should, before terminating it, return to its main topic and remind ourselves that the success of the Salam–Weinberg model is a great triumph for the basic ideas at the heart of contemporary theoretical physics.

§ 30 Symmetry breaking

This section deals with a phenomenon which apparently plays a large role in nature, which certainly occupies much space in current research journals and which is not at present well understood.

We have so far regarded the magnitude, at a particular point in space, of the wave ϕ associated with a particle as being a measure of the likelihood of finding the particle at that point. In so doing we have (implicitly) assumed that the lowest energy state, which corresponds to zero particle density, has $\phi = 0$ everywhere. For many Lagrangians this will be true, but it is possible to write down Lagrangians for which, in the lowest energy state, ϕ is not zero; instead, $\phi(\boldsymbol{r}, t) = C \neq 0$, where C is a constant (i.e. independent of position and time). This does not cause us any problems since we can simply define a new wave, say $\tilde{\phi}(\boldsymbol{r}, t)$, by

$$\tilde{\phi}(\boldsymbol{r}, t) = \phi(\boldsymbol{r}, t) - C \tag{30.1}$$

and rewrite the Lagrangian in terms of $\tilde{\phi}$ rather than ϕ. Then $\tilde{\phi} = 0$ would be the lowest energy (no-particle) state and we could proceed as usual. However—*it can happen that in doing this we have lost a symmetry of the Lagrangian*, i.e. we have *broken* a symmetry.

For a simple example, we consider two particles and their associated functions, ϕ_1 and ϕ_2, and suppose that the Lagrangian is symmetrical under transformations that mix these two fields. This means that it depends on quantities like

$$\phi^2 = \phi_1^2 + \phi_2^2 \tag{30.2}$$

which are unchanged by such mixing (cf the discussion following equations (5.1) and (5.2)). Now suppose we have a potential energy term in the Lagrangian which looks like the curve in figure 30.1. This is consistent with our symmetry, since it depends only on ϕ^2. However, it is clearly of the type discussed above, since its smallest value occurs not at $\phi^2 = 0$ but at

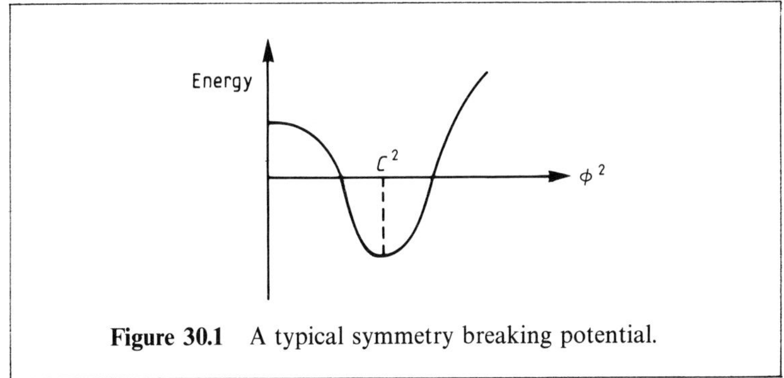

Figure 30.1 A typical symmetry breaking potential.

$$\phi^2 = C^2. \tag{30.3}$$

(Those who want an algebraic form might use $a - b\phi^2 + c\phi^4$, with a, b and c positive numbers. This has a lowest value of $a - b^2/4c$ at $\phi^2 = b/2c$.)

Of course, we now know how to overcome this little difficulty: we must define two new fields $\tilde{\phi}_1$ and $\tilde{\phi}_2$ by

$$\tilde{\phi}_1 = \phi_1 - C_1 \tag{30.4}$$

$$\tilde{\phi}_2 = \phi_2 - C_2 \tag{30.5}$$

such that when $\tilde{\phi}_1 = \tilde{\phi}_2 = 0$, ϕ_1 and ϕ_2 satisfy equation (30.3). Clearly, this requires

$$C_1^2 + C_2^2 = C^2. \tag{30.6}$$

However, this is only one equation, and it does not tell us enough to determine C_1 and C_2 separately. Well, we might say, we just make some suitable choice of C_1 and then let equation (30.6) give us C_2 (e.g. take $C_1 = 0$, then $C_2 = C$). In practice, this is what we do, but—and this is the key point—in so doing we spoil the original symmetry of the Lagrangian (the choice in the previous parenthesis is not symmetrical between 1 and 2). In general, a choice of C_1 and C_2 subject to equation (30.6) involves a choice of a *direction* relative to the ϕ_1, ϕ_2 axes (see figure 30.2) and the symmetry has gone. Thus, we have a symmetrical Lagrangian but non-symmetrical solutions!

Having fixed the vacuum we must look at the deviations from it, i.e. at the $\tilde{\phi}_1$ and $\tilde{\phi}_2$, in order to study the particles of the theory. Something interesting happens now. To appreciate this we look at

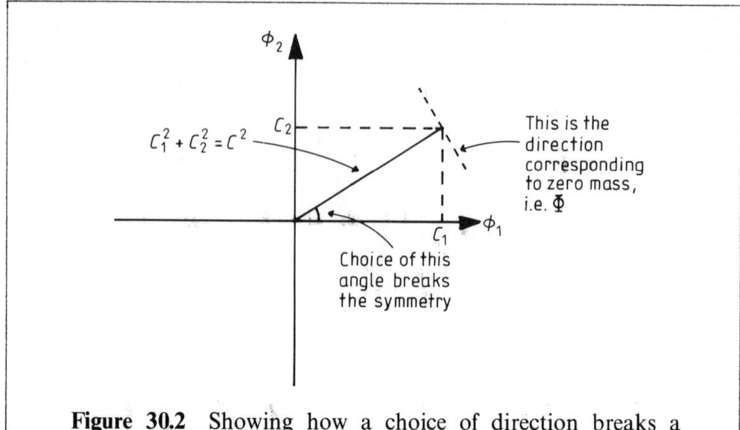

Figure 30.2 Showing how a choice of direction breaks a rotational symmetry. The vacuum state corresponds to *any* line of length C.

particular linear combinations of the $\tilde{\phi}_1$ and $\tilde{\phi}_2$ which take us either along the chosen direction in the ϕ_1, ϕ_2 diagram (figure 30.2), or perpendicular to this direction. In fact, the appropriate combinations are $\chi = C_1\tilde{\phi}_1 + C_2\tilde{\phi}_2$ and $\Phi = -C_2\tilde{\phi}_1 + C_1\tilde{\phi}_2$. For small values of the latter there is no change in the energy (because of the original symmetry), which implies that there is no Φ^2 term in the Lagrangian, i.e. that Φ corresponds to a *massless*, spin zero, particle. This is called a Goldstone boson and the existence of at least one such particle is a general feature of symmetry breaking.

What has all this to do with the real world? First, we have already stated (§28) that QCD has an (approximate) chiral symmetry which would be exact if the quarks had zero mass. Such a symmetry forbids composites to have mass, so it is contrary to experiment, and must therefore be broken. Some such mechanism as the above is generally believed to be responsible for this breaking. Nobody really knows how this happens and there is certainly a need for more understanding here. The pion is one of the Goldstone bosons and would therefore be massless if the u and d quarks had zero mass.

Of more immediate concern to us is the problem of breaking the gauge symmetry in the Salam–Weinberg model of the previous section. The breaking of a *local* symmetry, as we require there, is called the Higgs mechanism (after its discoverer). It has some remarkable, and very useful, properties. To understand these we note that the spin 1 particles ($W^{\pm}$, Z^0), introduced by the local invariance, must interact with the spin zero particles. The argument here is very similar to the discussion given in §21, where we saw that they had to interact with spin $\frac{1}{2}$ particles. There are, however, some important differences, due to the fact that the kinetic energy term in the Lagrangian contains the square of the spin zero field ($(\partial\phi/\partial x)^2$, rather than $\Psi(\partial\Psi/\partial x)$ as in equation (21.2)). In consequence, when this is made locally invariant we obtain terms which couple the square of the spin 1 fields to the spin zero fields. When the latter take a non-zero value in the lowest energy state, these terms become mass terms for the spin 1 fields. Thus, as we require, the $W^{\pm}$ and Z^0 acquire a mass. The way this happens is very pretty and removes another potential embarrassment. *Massless* spin 1 particles actually have one less degree of freedom than *massive* spin 1 particles; they can only have spin projections along their velocity of ± 1; zero is not allowed. Thus, when they acquire a mass we appear to gain an extra degree of freedom for our theory. This is not possible, so from where has the extra state come? The answer is that the massless spin zero particle (the would-be Goldstone boson) has disappeared and has become the extra degree of freedom of the spin 1 particle.

The details of all this are rather messy but, almost magically, it all works. The final Lagrangian is renormalisable. It also contains one or more spin zero particles (Higgs bosons) with masses which are expected to be around the $W^{\pm}$ and Z^0 masses. These correspond to the fields (χ) in the direction of the chosen axis in figure 30.2. Such particles should be seen in experiments and, indeed, there are already rumours of such observations at the CERN Collider.

Within the Higgs symmetry breaking scheme there is also a reasonably natural way of giving masses to the quarks and leptons. Unfortunately, each such mass requires an arbitrary constant, so no further tests of the theory are produced.

In constructing their successful model of weak interactions, Salam and Weinberg used the simplest possible Higgs breaking mechanism. This choice has, apparently, been vindicated by the confirmation of their predictions. Whether their Lagrangian represents the 'truth' at the fundamental level, or whether it is a low-energy approximation to some other theory, are questions, the discussion of which we defer to the final chapter, and the answers to which must await further experimental developments.

Summary of Chapter Three

We have met a new, weak, short-range interaction with several novel features (§§25, 26, 27). Attempts to understand this using the methods which worked so well in the previous chapters, i.e. by going to a new level of compositeness (§28), have met serious problems. However, the use of the Higgs symmetry breaking mechanism (§30) has allowed us to find a renormalisable Lagrangian in which there is a partial unification of the weak and electromagnetic interactions (§29). This model successfully predicted the existence and structure of the neutral current weak interaction, and the mass of the $W^{\pm}$ and the Z^0 (§29).

We have now arrived at what is known as the standard model (QCD plus the Salam–Weinberg theory of weak and electromagnetic interactions). Apart from gravity—which appears to be completely separate—and some small effects which we have not mentioned but can be incorporated quite easily (CP violation) we have covered all of physics. There is not a single, confirmed, experimental fact which goes beyond the standard model†.

Is *this*, then, at last, the end of our story? I believe the answer is no, but the reasons now are not experimental—they are aesthetic. This is

† We should mention here several unconfirmed possibilities. There are some recorded events which might be proton decay; we discuss these further in §31. There is also the curious fact, observed at the CERN $p\bar{p}$ Collider, that in three cases (about 30% of the total) the decay products of the Z^0 (i.e. e^+e^- or $\mu^+\mu^-$) are seen together with a high energy photon. The standard model cannot account for this photon and, indeed, it is hard to find any modification which can explain it, although it would be considerably more likely within the framework of the composite models of §28. Finally, in the same experiments, some 'events' which are unexpected and not readily explicable within the standard model have been seen. Further experimental data, which are being taken at the present time (end of 1984) should reveal whether these really are indications of new physics. A proper analysis of these data is likely to take many months.

just not a good way to end such a great story. In the next chapter we shall see why; we shall look for ways to a better ending, although we shall not be successful, and we shall explain why it *may* take experiments a long time to catch up!

Chapter Four

in which we leave experiments, with only good sense to guide us,
we learn about gravity,
we review some speculations,
we try to understand the Universe,
and close with some very interesting questions.

§31 Beyond the standard model

Why? What are the reasons for our dissatisfaction with this model as the solution to essentially everything in physics? In one sense it is the *success* of what we have done so far that makes us not content.

Readers will have already noticed that, whereas in the first two chapters we obtained, at least in principle, a lot of physics from a few general principles and one or two fundamental constants, this was not the case in the last chapter. Even though, at the start, there was actually not much that required explaining, we had to introduce several new, and somewhat ugly, constructions to get even as far as we did. The masses of the Higgs particles and of all the spin $\frac{1}{2}$ states, together with the mixings involved in d', s' and b' are input parameters, i.e. not, even in principle, calculable. Furthermore we saw no hint of any reason for the replication of families, or for the violation of parity in weak interactions.

Can we do better? An obvious way forward comes from noticing that the theory of electromagnetism was developed through a unification of electric and magnetic forces (once thought to be separate phenomena); similarly, the Salam–Weinberg model involved a unification of this theory with weak interactions. Surely the next step must be to somehow combine this with QCD, and to make a 'Grand Unified Theory', or GUT. The big difference between the strengths of the interactions would be an immediate deterrence to such a venture, except that we now know of two things that might help us. First, there is the possibility of symmetry breaking and, second, coupling constants vary with energy.

Our aim, then, is to find a new global symmetry, which we can make into a local symmetry (by the standard procedure), such that some of the spin 1 particles which we introduce are the gluons, some are the $W^{\pm}$, one is the Z^0, one the photon, and there may be others. In fact, we do not need to search very far, because such a symmetry is already apparent. Our Lagrangian contains quarks and leptons and we have

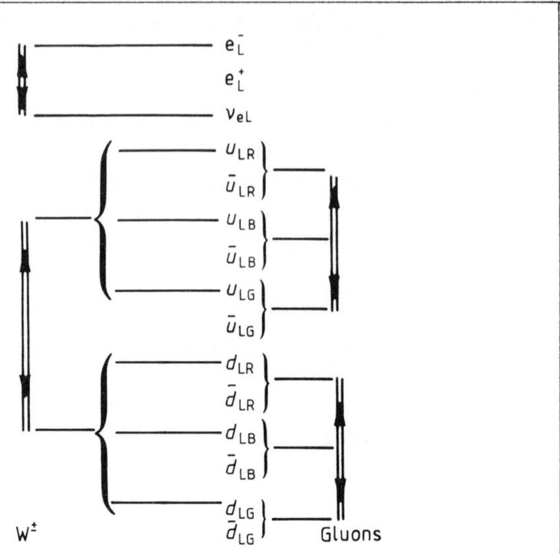

Figure 31.1 A single family of left-hand fermions. The Lagrangian containing just these fermions, with no mass terms, has a global symmetry associated with all possible mixings of the 15 states. On the left-hand side of the picture we see how the $W^{\pm}$ are associated with simultaneous mixing of u states with d states and e^- states with ν_{eL}. Similarly, on the right-hand side we see how the gluons are associated with simultaneous mixings among the three colours.

not yet exploited the symmetry involved in mixing these. To be explicit, we show in figure 31.1 a single family. In this figure we have for convenience just included the left-hand states, i.e. instead of putting in u_L and u_R, for example, we have used u_L and $\bar{u}_L$. When we also add the antiparticle set of states (which in this case will all be right-handed) we will, of course, have everything. We have not put in a state $\bar{\nu}_{eL}$ (or ν_{eR}) since it is not necessary (unless the neutrino turns out to have a mass). In the figure we have attempted to show the symmetries that have already been used. Clearly, however, there are many more. Indeed we could consider arbitrary mixings of all 15 states, which would introduce $15\times15-1=224$ massless vector particles. In fact, for technical reasons, we are forced to something rather less elegant, namely we take five states (e_L^+, ν_{eL}, $u_L^{\rm red}$, $u_L^{\rm green}$, $u_L^{\rm blue}$)

and consider all possible mixings of these, together with a suitable selection of mixings of the other ten†. The number of spin 1 particles to be introduced when we make this global invariance into a local invariance is 24 ($=5\times5-1$; cf §23). Eight of these will be massless gluons, one will be the photon, three will be the $W^{\pm}$ and Z^0; the remaining twelve are *new* and give rise to processes we have not met, e.g. figure 31.2, in which X is one of the 12 new particles. Since such interactions are not seen, the symmetry must be broken so that the X particles become heavy. How heavy?

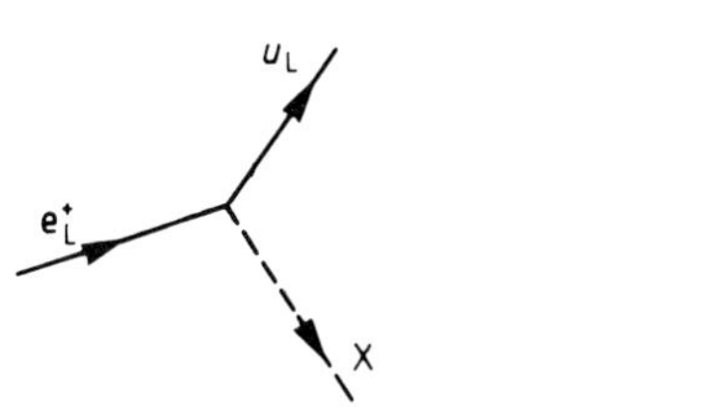

Figure 31.2 A new type of interaction introduced in the Grand Unified Theory.

Before answering this question, we do a small calculation. Let M_X be the mass scale associated with the Higgs mechanism that we will have to introduce to break our new symmetry. This will be approximately the same as the mass of the X particle of figure 31.2. Then at energies much greater than M_X the QCD coupling constant and the Salam–Weinberg couplings will be equal (because a gauge theory only allows *one* coupling‡). However, at lower energies the symmetry breaking becomes significant and these couplings vary independently. In fact they vary at a different rate because they are associated with mixing between 3 objects (QCD), between 2, or with a

† In group theory language we do not gauge SU(15), but SU(5), in which we have the *5* and $\overline{10}$ representations.

‡ Readers with a good memory will object that we allowed two in §29. This was possible because there we gauged both the 'mixing' symmetry and the equal-phase-change invariance. The latter is separate and has an independent coupling. Here, since we do not *need* to introduce it, we don't. In technical language, this means that we make SU(5) local; not U(5).

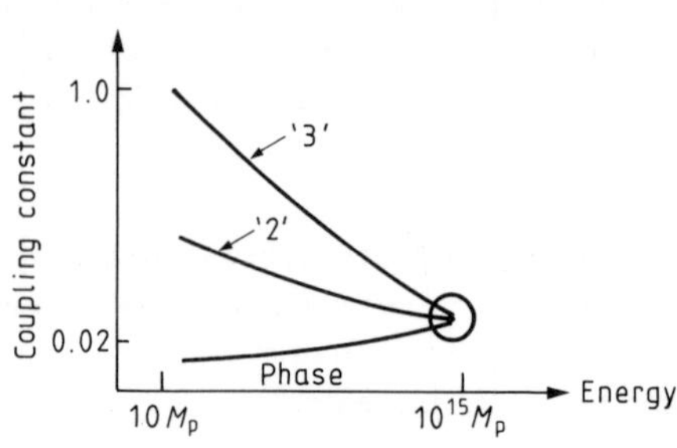

Figure 31.3 Showing how the QCD, weak and electromagnetic coupling constants vary with energy from being equal at about $10^{15} M_p$ to their observed values at low energy. (Note that the nature of the energy variation is such that this graph is drawn with the horizontal axis as the logarithm of the energy, rather than the energy itself.)

phase (Salam–Weinberg). The way they vary can be calculated and, from knowledge of the values at low energy, we can estimate where—and if—they come together. The calculation yields curves that are depicted in figure 31.3.

This picture is very encouraging. The curves all meet at one point (they didn't *have* to) and, as we require, M_X is very large (about 10^{15} M_p). Is it large enough? To answer this we must consider the most significant effect of X exchange, namely proton decay:

$$p \rightarrow \pi^0 + e^+ \tag{31.1}$$

which takes place as shown in figure 31.4.

Since protons are well known to be 'stable' (by which we mean that, if they have a lifetime, it is enormously long), the whole idea of GUTS could fail disastrously at this point. However, because the value of M_X is so large, the immediate disaster is avoided. The proton lifetime, due to the decay in figure 31.4, is calculated to be about 10^{30} years. We stress that this really is a long time! If we had some protons at the start of the Universe then we would need to *multiply* the present age of the Universe by a factor 10^{20} in order for half of them to have decayed.

Naturally the above prediction was a challenge to experimentalists and, fortuitously, it is just within the limits of possible experimental

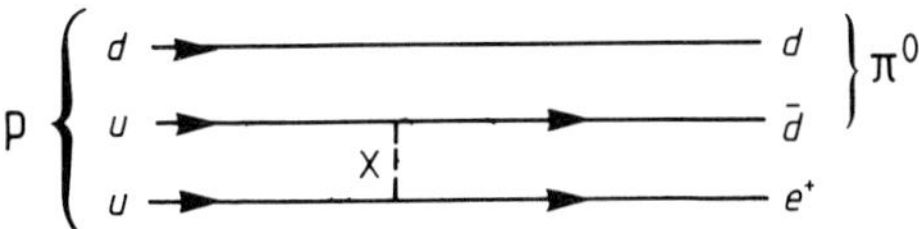

Figure 31.4 Showing how the X particle introduced through the GUT can lead to decay of the proton into a pion and a positron.

observation. Basically the experiments consist in burying some material deep in the Earth, to eliminate cosmic rays, surrounding it with detectors and then waiting to see if anything happens. Since it is not possible to shield the apparatus from neutrinos, the detectors have to measure the event sufficiently carefully for proton decay and neutrino interactions to be distinguished. The experiment under Mont Blanc, referred to in §1, has reported a few events which appear *not* to be caused by neutrino interactions. If they are proton decays then they would correspond to a lifetime of about the predicted value. However, a group working with 10^4 tonnes of water, in a salt mine 600 m below Lake Erie in the USA, have seen no evidence for proton decay and have thereby established a lifetime for the process (31.1) which is greater than 2×10^{32} years.

Although there are some uncertainties in the theoretical calculations, this result appears to rule out the *simplest* form of GUT as described earlier. Modifications are possible and these have the right effect, i.e. they increase the predicted value of the proton lifetime. As we shall see later, it may even be possible to explain the apparent discrepancy between the protons on the French/Italian border and those near Chicago.

If the proton lifetime is much longer than 10^{32} years, then it is unlikely that any experiment will ever see it. Where then would we have to look for evidence of anything outside the standard model? One possibility is that, in the modified forms of GUT, with longer proton lifetimes, there are likely to be other particles with masses in the region between a hundred proton masses (the M_W scale) and 10^{15} M_p (the M_X scale) and some of these may be observable in the near future. Nevertheless, real evidence for GUTs might be hard to obtain and, in such a situation, we should look closely at its aesthetic appeal.

On this basis it must be admitted that the GUT which we have discussed is not really very attractive. In spite of the encouraging successes noted above (and some others), it does not solve any of the problems with which we started this section. The number of input parameters and assumptions is *increased*, rather than decreased, and we have not discovered any reason for the presence of the three (at least) families. In addition, the GUT introduces a serious new difficulty: *the hierarchy problem.* In its simplest form this asks why there should be two such widely different mass scales in physics, $10^2\ M_p$ and $10^{15}\ M_p$, but it is actually more serious than this. We might be content to regard these two scales as input to some original Lagrangian, but we would then have to worry about finite renormalisation corrections which would tend to mix the two scales. In particular, the light particles would receive 'corrections' to their masses which would be roughly equal to the masses of the heavy particles, so the two scales would effectively disappear and all particles would be heavy. This could only be prevented if there were miraculous cancellations between various contributions to some of the particle masses. For such cancellations to be sufficient it would be necessary that certain input parameters were adjusted to about 1 part in 10^{14}. This could have occurred but it is very unnatural, and it gives rise to an uneasy feeling that, unless the whole idea of GUT is a mistake, there is something happening which we have not yet understood. Later, we shall meet other, similar, 'fine-tuning' problems.

To conclude this section: we are encouraged by the successes of GUTs to believe that they may well contain much that is truc, although we would be much happier if we had decisive evidence for proton decay. There is naturally a lot of activity in trying to improve the available GUTs and to solve some of their problems. Before we discuss this activity we shall, at last, digress into gravity.

§ 32 Gravity

There is no natural place in our story for this section, which fact is an indication that gravity does not fit comfortably into our understanding of fundamental physics. Maybe it really is separate from the other forces; more likely is the possibility that these other forces cannot be completely understood until it has been correctly incorporated.

Gravity is the weakest, though the most immediately evident, of known forces. It causes objects to drop. They drop because there is an attractive force between them and the Earth. Perhaps the greatest single step ever made in theoretical physics was that made by Newton when he showed that a simple law of gravitational attraction

$$\boldsymbol{F} = -G\frac{m_1 m_2}{r^2}\hat{\boldsymbol{r}} \tag{32.1}$$

correctly predicted the properties of planetary orbits, thus bringing the 'heavens' within the scope of human understanding.

This law, as we already noted in §7, has exactly the same form as that giving the electric force. Instead of electric charges, however, the gravity force involves the masses, m_1 and m_2, of the particles. This fact is sometimes referred to as the equality of 'inertial' mass, which appears in equation (7.1), and 'gravitational' mass, which appears in equation (32.1). An immediate consequence of this equality can be seen by calculating the accelerations of various particles due to the gravitational force of some fixed object, e.g. the Earth. Then the masses of the particles cancel and *equal* accelerations are obtained; hence the result of Galileo's famous, though perhaps mythical, experiment, in which he demonstrated that different objects take the same time to reach the ground when dropped from Pisa's leaning tower (only true, of course, to the extent that air resistance effects are the same).

Since the mass of an object is a positive number, a second

consequence of equation (32.1) is that the force of gravity is always attractive. This is the reason why gravity is so important in our environment. Whereas the electric charges of protons and electrons cancel, so that the Earth is (approximately) electrically neutral, all particles within it contribute with the same sign for the gravitational force, which is therefore very evident.

The number G, the gravitational constant, can be seen from equation (32.1) to have units $(\text{mass})^{-1}$ $(\text{length})^3$ $(\text{time})^{-2}$. Its value is determined experimentally to be

$$G = 6.672 \times 10^{-11}\ \text{m}^3\,\text{kg}^{-1}\,\text{s}^{-2}. \tag{32.2}$$

This is a new fundamental constant of nature.

It is not possible to make a pure number (i.e. a quantity without units) out of G and the other fundamental constants h and c. We can however make a mass, which is called the *Planck mass*, defined by

$$M_{\text{PL}}^2 = (hc/2\pi G). \tag{32.3}$$

This has the approximate value

$$M_{\text{PL}} = 10^{19}\,M_{\text{p}}. \tag{32.4}$$

We can also make a time, the *Planck time*, given by

$$T_{\text{PL}}^2 = \frac{h}{2\pi}\frac{G}{c^5} \tag{32.5}$$

with value

$$T_{\text{PL}} = 10^{-43}\ \text{s}. \tag{32.6}$$

The magnitudes given in equations (32.4) and (32.6), which will reappear later in our story, are indications of the weakness of gravity, i.e. the fact that G is small. Another demonstration of this is the ratio of the electric and gravitational forces between an electron and a proton:

$$\frac{e^2}{GM_{\text{p}}m_{\text{e}}} = 2.3 \times 10^{39}. \tag{32.7}$$

This number clearly justifies the neglect of gravity in atomic physics.

We must now make a proper theory out of gravity, consistent with special relativity, analogous to the development of electromagnetism from equation (7.3). This leads us to general relativity, a subject often regarded as very mysterious and impossibly difficult. This is not really fair; the basic idea (like all great ideas?) is beautifully simple.

In our introduction to special relativity we noted that absolute velocities had no significance; only relative velocities could be detected. At first sight, however, this is not true for accelerations. We cannot detect the velocity if we sit in a train with the blinds down, but we can certainly detect acceleration; we can *feel* when the train is stopping. Einstein was unhappy about this; why should not observers see the same physics even when they are in relative acceleration? He realised that an acceleration of the observer has the effect of giving all objects an equal acceleration, i.e. it is equivalent to applying a gravitational force. In other words, accelerating observers and gravitational fields are the same thing. This fact, which is a consequence of the equality between gravitational and inertial masses, should somehow be built into the theory. Now the description of the motion of an observer is really 'geometry' (in the 4-dimensional space t, $\boldsymbol{x}$); so we are led to the idea that gravity can be understood *geometrically*.

To this end, consider the motion of a particle under no forces. Its path is a straight line† in the 4-dimensional space (this is equivalent to saying that it *moves with constant velocity* along a straight line in ordinary 3-dimensional space). We would like to be able to make a similar statement even when gravitational fields are present. This is possible if we have a new *definition* of what we mean by a 'straight' line. This definition must be such that, for example, the elliptic orbits of planets are straight lines. Now the simplest definition of a standard straight line is that it is the shortest distance between two points. Thus, we can generalise the idea of a straight line if we allow alternative definitions of distance. These are easy to find. Consider again the distance from an origin O to a point P, as already discussed in §5. Instead of writing for the square of this distance, OP^2, the expression x^2+y^2 given by equation (5.3), we use something more general, in which the x^2, etc, terms are multiplied by different numbers, and 'cross-terms', involving xy etc, are included. Actually, since we wish to bring time into our discussion, it is the 4-dimensional length, defined at the end of §15, which we must generalise. Thus, we define the 'distance', s, between the space–time point $(0, 0, 0, 0)$ and

† See, for example, the path of the point P depicted in figure 15.2.

the space–time point (x, y, z, t) as

$$\begin{aligned} s^2 = g_{00}t^2 &+ 2g_{01}tx + 2g_{02}ty + 2g_{03}tz \\ &+ g_{11}x^2 + 2g_{12}xy + 2g_{13}xz \\ &+ g_{22}y^2 + 2g_{23}yz \\ &+ g_{33}z^2. \end{aligned} \tag{32.9}$$

Here we have introduced the numbers g_{00}, g_{01}, g_{11} etc. Collectively these are called the 'metric'. Note that it is purely a matter of convention that we have multiplied the cross-terms, e.g. $g_{01}tx$, by 2.

We include all the effects of the gravitational force in the metric. Then, even when such forces are present, particles will travel along straight lines as defined by this metric. Readers who are objecting that, although this might be a very elegant idea, it is completely empty until we have given some rules for finding the metric, are of course quite correct. Before we discuss this, however, we shall explain some other simple, physical situations where a metric is involved. (Those who do not find these examples simple should omit the next three paragraphs.)

Suppose we have a string lying on a plane whose points are described by components x, y as in §5; see figure 32.1. The string clearly determines a value of y for each x; that is, it expresses y as a function of x: $y = f(x)$. We choose the origin so that the string passes through it (thus, $0 = f(0)$), and consider the length of the string from $(0, 0)$ to a nearby point (x, y). This is clearly given by $l^2 = x^2 + y^2$,

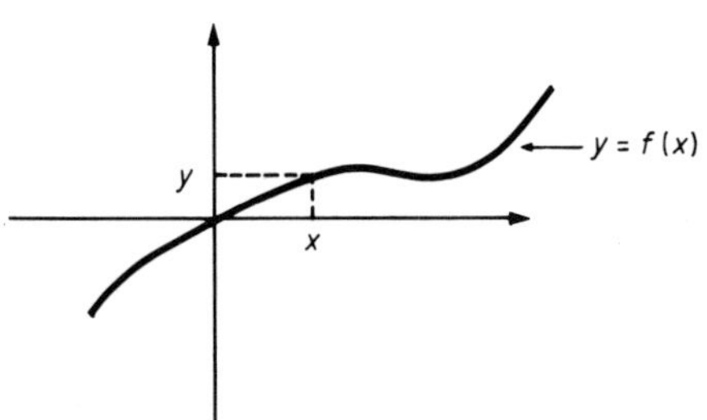

Figure 32.1 The heavy line gives a value of y for each of x. For small x the value of y is equal to the product of x and the 'slope', which is written as $\partial y/\partial x$.

provided x and y are sufficiently small that the string can be regarded as straight. Since (x, y) is on the string we can write this as $l^2 = x^2 + f(x)^2$. However, for small x, we can put

$$f(x) = f(0) + x\frac{\partial f}{\partial x} = x\frac{\partial f}{\partial x} \tag{32.9}$$

(compare equation (21.1); this equation defines the 'derivative' function $\partial f/\partial x$). Then

$$l^2 = \left[1 + \left(\frac{\partial f}{\partial x}\right)^2\right]x^2 \tag{32.10}$$

which we can write in the form

$$l^2 = gx^2 \tag{32.11}$$

with g being the expression in square brackets in equation (32.10). We see that equation (32.11) is the one-dimensional analogue of equation (32.8). Note that for a straight string we could eliminate the need for g (i.e. put $g = 1$) by simply choosing the x axis to be along the string.

For a two-dimensional analogue we could think of the distance between two points on the surface of a wine glass, described now by $z = f(x, y)$. Again, if we choose the origin as one of the points we have, for small x, y and z,

$$\begin{aligned} l^2 &= x^2 + y^2 + z^2 \\ &= x^2 + y^2 + f(x, y)^2 \\ &= x^2 + y^2 + \left(x\frac{\partial f}{\partial x} + y\frac{\partial f}{\partial y}\right)^2 \\ &= \left[1 + \left(\frac{\partial f}{\partial x}\right)^2\right]x^2 + \left[1 + \left(\frac{\partial f}{\partial y}\right)^2\right]y^2 + 2\left(\frac{\partial f}{\partial x}\right)\left(\frac{\partial f}{\partial y}\right)xy \end{aligned} \tag{32.12}$$

which has the required general form. Here again, if the surface were flat, we could choose the axes so that the metric would reduce to that of equation (5.1). Thus, the need for a different metric arises because of curvature. It is in this sense that we say that gravity 'curves' space; only when the field of gravity is zero is it possible to choose coordinates so that equation (5.3) gives the distance. Unlike the examples of the string and the wine glass, we cannot 'see' the curvature of our 4-dimensional space because to do so we would need to step outside it into a space of higher dimensions (compare the string in a plane, or the surface in 3-dimensional space).

It is important to realise here that curvature is an intrinsic property of a particular space and does not *require* the space to exist in higher dimensions. We can illustrate this by considering an ant on the surface of the Earth. In fact, we require a somewhat idealised ant that cannot look upwards and has zero height, so that it lives entirely in two space dimensions and is completely unaware of the existence of a third. In spite of this it can observe that the Earth's surface, i.e. the space in which it lives, is curved. For example it could draw triangles. (Here the reader should become our ant and draw a triangle, e.g. figure 32.2.) On adding up the internal angles it would find 180 degrees (see figure 32.2). From this fact it would deduce that the

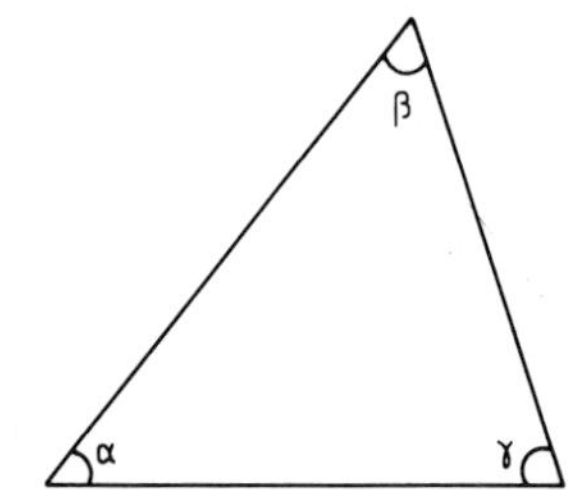

Figure 32.2 A triangle on a plane. The angles α, β, γ will always satisfy $\alpha+\beta+\gamma-180°$.

surface is nearly flat, i.e. the curvature is small. However, it would not obtain *exactly* 180 degrees, and the deviation would increase as the triangles became larger. To show this, we take a very large triangle. We suppose the ant begins on the equator at São Joachim (Brazil) and moves east until it reaches Befori (Zaire), by which time it has travelled one-quarter of the distance around the Earth. Then it turns through 90 degrees and walks north, until it reaches the North Pole. It now turns again through 90 degrees and walks south. After a further 8000 miles it will return to its starting point, at 90 degrees to its original direction. On a plane the path would be that shown in figure 32.3. In fact however, as shown in figure 32.4, the ant will have walked

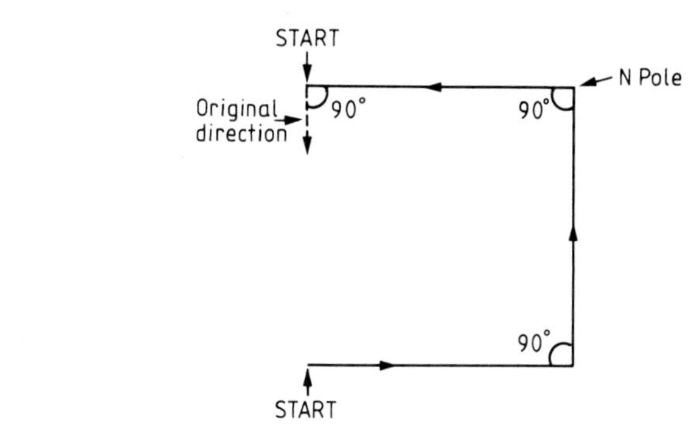

Figure 32.3 Showing the path of an ant on the surface of the sphere. If each line has a length equal to a quarter of the circumference then the path will in fact form a triangle. (Check this on a suitable sphere—or use figure 32.4.)

in a triangle, whose internal angles add to 270 degrees. If the ant has only heard of geometry on a plane, then it will be surprised. A better educated ant will realise that it lives on a curved surface, and it will even be able to calculate the curvature.

We now return to gravity; in particular, to the metric. This must contain all the information about the gravitational force, so we expect, because of equation (32.1), that it will be related to the masses.

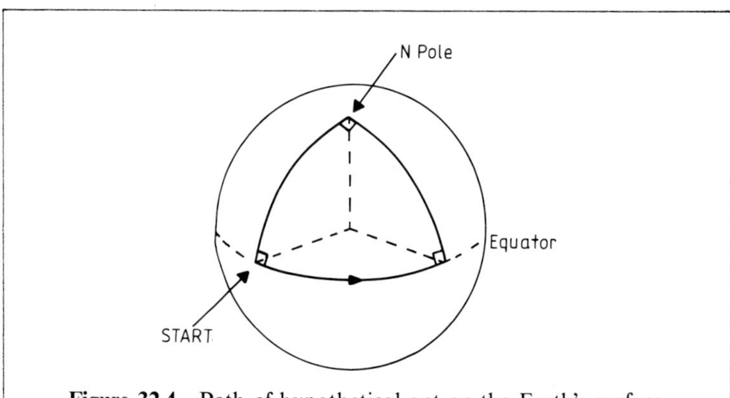

Figure 32.4 Path of hypothetical ant on the Earth's surface.

Einstein obtained the correct relation by requiring that it was as simple as possible, consistent with special relativity and, of course, in approximate agreement with the tested law of gravity (equation 32.1)) for suitable circumstances ('weak' fields). We shall not write Einstein's equations here, but merely state that they relate the metric to the source of gravity, which is the mass and energy distribution in space–time. They define the theory known as general relativity. Empty space is 'flat', i.e. has $g_{00} = -g_{11} = -g_{22} = -g_{33} = 1$ and the other gs equal to zero. In general, the $g_{\mu\nu}$ (where μ, ν can be any of 0, 1, 2 or 3) will depend upon position and time. This implies that we should only use equation (32.8) if the points are sufficiently close so that $g_{\mu\nu}$ does not vary greatly along the line joining them.

The equations of general relativity give some small corrections to the weak-field approximation (equation (32.1)), in particular to the planetary orbits. Naturally, the experimental observation of these corrections gave a big boost to the theory. Away from the regions where the weak field approximation can be used as a starting point, i.e. where gravity is in some sense a small correction, there is no complete understanding of all the possible solutions of Einstein's equations for the metric (they are non-linear and hence much more difficult to treat than, for example, the equations of electromagnetism). Some solutions have 'singularities', of which 'black holes' are the best known example.

It is possible to interpret the metric $g_{\mu\nu}$ as a quantum-mechanical wave associated with a spin 2 particle (the spin is deduced from the fact that there are two indices, μ and ν), which is called the 'graviton'. Thus the force of gravity, equation (32.1), can be seen as the exchange of a *massless* graviton. It is of course massless because the force, like the electric force, has infinite range. By analogy with electromagnetic waves, e.g. light, which are beams of photons, there should exist gravitational waves, which are beams of gravitons. Because of the weakness of the coupling it will require detectors of incredible sensitivity to detect these waves, so it is not surprising that attempts so far have not been successful.

The Lagrangian describing spin 2 particles is not renormalisable, which is one reason why there is no satisfactory quantum theory of gravity. The theory is not a local gauge theory in the previous sense (since such theories give spin 1 fields), but it can be derived on similar lines. The argument, briefly, is as follows. The equivalence of gravitational fields and accelerations means that 'locally' (i.e. at a fixed place and time) one can choose an observer who does not see a gravitational field (to an observer in a freely falling lift, for example, everything in the lift appears to be moving with uniform velocity—this is a 'thought' experiment; don't try it!). This freedom to choose, locally, such an observer is somewhat analogous to the freedom to choose a phase, for example, locally. Hence we see that there are aspects of gravity that make it similar to a gauge theory; we repeat, however, that it is not renormalisable.

At this point we recall the problem of absolute acceleration, noted below equation (32.7), and must admit that we have not solved it. The equivalence between an accelerating system of reference and a gravitational field, which is a key feature of general relativity, does not of course explain either. It is still true that non-accelerating systems are special; so the question remains: with respect to what system is this acceleration measured, or, in other words, what defines zero acceleration? A possible answer might be the system obtained by averaging in some way over distant matter (i.e. the galaxies—see §34). If this is the case, then inertial mass—and hence also gravitational mass—have their origin in such distant matter. This idea, which can be expressed in many ways, is known as Mach's principle. It was proposed before general relativity and has always been a subject of controversy. Its status remains unclear.

Finally, we must discuss the 'cosmological constant'. This is a number which Einstein realised could be added to his equations relating $g_{\mu\nu}$ to the distribution of matter. It actually represents a constant background energy density and would be unobservable apart from its effect here. Einstein was originally motivated to include a non-zero cosmological constant because he realised that, without it, the simplest solutions of the equations predicted an expanding Universe, which, when extrapolated back in time, led to a 'starting-time' or moment of creation. Such a concept was against the prevailing fashion†, so the constant was introduced to prevent it. As we shall see in §35, our prejudices have now changed.

† It has been suggested—see, for example, the article by Wheeler in *Some Strangeness in the Proportion* (New York: Addison–Wesley 1981)—that Einstein was much influenced here by Spinoza who, almost three centuries earlier, had rejected the idea of creation on the ground that before anything existed there could not be a clock to tell creation when to begin. I leave the reader to make his own assessment of the validity of this argument.

Present observations are consistent with the cosmological constant being zero. Certainly it is 'small', by which we mean that it is *very much* smaller than effects which we now know can contribute to it. For example, the difference in energy between the symmetric and non-symmetric states of the GUT theories discussed in the previous section contribute to the cosmological constant. If we suppose (reasonably) that at some early time the high temperature of the Universe (see §35) ensured that it was in the symmetric state, then, when it cooled and jumped into the observed non-symmetric state, the cosmological constant would change by more than 10^{50} times its present value. Thus, this change must be cancelled, to an accuracy of 1 part in 10^{50}, by the previous value. Other contributions to the cosmological constant appear to require cancellation to an accuracy of the order of 1 part in 10^{120}. This is an even more acute fine-tuning problem than the one we met in §31. No convincing solution to this problem is known.

We shall mention gravity again later, but, for the moment, we return to have a last look at theories of elementary particles.

§ 33 Super theories?

Although we have reached the end of our discussion of topics that have obvious relevance to observational physics, lots of questions remain. These lead us into speculations that *might* be relevant to physics, and might even answer some of the questions.

Composite models of quarks and leptons, already discussed in §28, may be the way forward, but at present it seems that they need some new ideas. Theories in which the Higgs particles are composite (but not the quarks and leptons) so that their properties could be *calculated*, with a consequent reduction in the arbitrariness, were popular a few years ago. Since the likely composites of quarks are already known, i.e. the familiar states of 'low-energy' particle physics like the pion and the proton, it was necessary to introduce a new set of heavier 'quarks' bound by a new QCD-like interaction ('techni-quarks' with 'techni-colour'). The theories became increasingly complex and unnatural, in an endeavour to explain the facts, and appear to have died.

The idea that is attracting most attention at the present time is 'supersymmetry', although in the words of one of its inventors, B Zumino, 'considering that there is no experimental evidence whatsoever that supersymmetry is relevant to the world of elementary particles, it is remarkable that there is so much interest in the ideas'†.

In supersymmetric theories the Lagrangian is constructed so that it is invariant with respect to interchanges of particles with different spin, for example, spin 0 with spin $\frac{1}{2}$, or spin $\frac{1}{2}$ with spin 1. That this is not an 'innocent' extension of previous ideas can be seen from the following argument. Recall that for a spin $\frac{1}{2}$ particle the kinetic energy term in the Lagrangian has units ψ^2 divided by length (see for example equation (21.2)). However, for a spin 0 or spin 1 particle, the

† *CERN Courier* **23** 18 (1983).

corresponding term† is ϕ^2 divided by (length)^2. Since all terms in the Lagrangian have the same units (in fact (length)$^{-4}$) it follows that the units of ψ (spin $\frac{1}{2}$) and ϕ (spins 0, 1) are different: ψ is (length)$^{-3/2}$, ϕ is (length)$^{-1}$. Thus we cannot just mix them as we did with colours in §23 (see below equation (23.1) for example). In fact, it is necessary that the mixed states involve derivatives ($\partial\phi/\partial x$) which implies that space and time are automatically involved in supersymmetry. This is an important new feature.

Does this development have any relevance to physics? It clearly requires extra particles; for example, if we have some Higgs spin 0 particles in a supersymmetric theory, then there will inevitably have to be spin $\frac{1}{2}$ Higgs particles in addition; similarly, there will have to be supersymmetric partners of the quarks and leptons ('squarks' and 'sleptons'), and of the gauge vector particles ('gluinos', 'w-inos', 'z-inos', 'photinos'). This increase of particles produces an effect for which there may already be some (tiny) evidence, namely it reduces the rate of variation of coupling constants with energy. Hence the unification mass (M_X) in figure 31.3 becomes larger. This in turn increases the proton lifetime—which is good. However, if the Mont Blanc events (§31) are real, it might be possible to explain them. We use the fact that mass estimates are valid to within a possible factor of about 10–100. Thus, although the Higgs mass associated with the breaking of the symmetry at M_X must be around M_X, it *could* be smaller, in which case proton decay might proceed through exchange of the Higgs particle, rather than the X as in figure 31.4. The Higgs particle, however, has spin 0, so an effect rather like that in K decay (§27) occurs, i.e. the interaction rate will contain the quark mass as a factor. This would strongly favour processes involving the *s* quark, i.e.

$$\mathrm{p} \to \mathrm{K}^+ + \nu \qquad (33.3)$$

rather than the process in equation (31.1). Such an event would not be seen by the detectors in the American experiment, but it would in the Mont Blanc experiment, so it could be the reason for the discrepancy between the results of these two proton decay experiments.

There are several general problems to which supersymmetry might provide solutions. It could, for example, help to keep certain spin 0 or spin 1 particles massless, by linking them with spin $\frac{1}{2}$ particles that *have* to be massless because of the chiral symmetry (§28). Such a

† As already noted in §30 it has the form $(\partial\phi/\partial x)^2$.

mechanism could help with the hierarchy problem of §31, or with the problem of light spin 1 particles in composite models (§28). The small observed masses would then be due to breaking of the chiral symmetry. In addition the extra constraints involved in supersymmetry reduce, to some extent, the excess freedom involved in symmetry breaking mechanisms. Several possible models are available. They differ in the nature of the mechanism that breaks the supersymmetry. None are really convincing, but they tend to predict lots of new particles (as indicated above), with masses generally in the region of M_W. Clearly these are coming within observational limits, so some may soon be seen† . . .?

Undoubtedly the most exciting recent development is the extension of supersymmetry into a *local* symmetry. Because, as we saw above, the global symmetry involves space and time, it is inevitable that when we make the symmetry local we are involved with the way space changes from point to point, which is the subject of general relativity. Hence local supersymmetry involves gravity, for which reason it is called 'supergravity'. This is an encouraging development and may lead the way to a unification of gravity with the other forces of nature. In this connection it is worth noting that M_X of §31 (10^{15} M_p) is already close to the Planck mass of §32 (10^{19} M_p); maybe this is trying to tell us that the continuation of our story is impossible without gravity.

Although the general ideas of supergravity are quite simple, they lead to enormous algebraic complexity, and there are many as yet unanswered questions. The most important of these deals with the question of whether supergravity theories are finite. It is easy to see that they are not 'renormalisable' in the sense of §20, so, if they produce infinities, they must be wrong, since the infinities cannot be removed by renormalisation. However there are indications that they may not *need* any infinite renormalisations because all observable quantities will, when correctly calculated, be finite. It seems likely that this exciting possibility will be confirmed—or refuted—in the next few years. Another interesting question is that of uniqueness; does the requirement of local supersymmetry, together with other plausible assumptions, yield a unique Lagrangian? Here there are some

† The fact that nature seems to have arranged things so that, among the particles observed so far, none are supersymmetric partners of each other is already somewhat disturbing for believers in supersymmetry.

promising indications; in particular, the model with the maximum possible amount of local supersymmetry ('$N=8$ supergravity') seems to have a unique Lagrangian. Unfortunately this does not, at least in any simple way, correspond to the world as we know it. Maybe if we could 'solve' the theory, i.e. find its composite states, we would obtain something more like the real world, but, at this stage, nobody knows.

Many forms of supergravity appear more natural if expressed in more than the familiar three space dimensions. For example, the $N=8$ supergravity theory fits easily into a world of 10 space plus 1 time dimensions. It could be that our world takes the form it does because it really exists in a higher dimensional space. How then could we be not aware of this? To be specific, suppose that there are really four space dimensions; why are we so convinced that there are only three? Why cannot we move in this fourth direction? Quantum theory suggests a possible answer. We assume that, for some reason, the extra dimension is rolled up into a circle of small radius. Provided this radius is smaller than the inverse of any 'reasonable' mass/energy scale with which we are familiar, the uncertainty relation will ensure that we cannot fix positions more accurately, in this extra dimension, than the circumference of the circle. Thus, we, and all the things we observe, are *everywhere* on the circle, so we are completely unable to distinguish separate points on it. Only with extremely high energy probes (greater than hc/R, where R is the radius of the circle) would it become possible to 'see' the extra dimension.

We are now in the realm of today's research, which will become the theories (or the forgotten foolish ideas) of tomorrow. There is much speculation and this is likely to continue because, as we noted earlier, there are no experiments, i.e. we already have 'explanations' for all present observations. It may need some really new and unexpected experimental result to tell us the most fruitful direction in which to speculate. Maybe the CERN Collider is already showing the first indications of such surprises—as mentioned in the footnote on page 138; otherwise we may have to await experiments from LEP. There is, of course, the possibility that nothing new is seen, i.e. the standard model continues to work. This would be such a disaster for theoretical physics that we shall think about it no further! Rather, until such time as the experimentalists give us something new, we shall have a closer look at the world as we know it.

§34 The Universe

Our story has taken us into the very small constituents of matter. Now we must change direction and consider the Universe. Knowledge of particle theory alone will never fully explain all observed phenomena; the performance is not wholly determined by the performers; we must know about the stage, the choreography—maybe even the plot.

Though our Universe is indeed made of quarks, leptons and photons, etc, it is surely not the *only* Universe which could be so constructed; why is it not bigger/smaller, more dense/less dense, more 'lumpy'/more smooth, etc? Can we explain any of its observable properties? It may be that considerations of the Universe, and how it came to have its present form, will restrict possible theories of the elementary particles from which it is made; similarly, what we know about those particles may help in understanding the structure of the Universe.

There is one very obvious reason why our story has reached the stage where we should look at the Universe. We have been led to distances that correspond to energies greater than about $10^{15}\,M_p$. Now LEP, the machine mentioned in §1, will give us about $200\,M_p$. Maybe, one day, we will build a machine† that goes up to $10^3\,M_p$, but there appears to be little hope of ever going significantly beyond that, so experiments at higher energies will (perhaps) always be impossible. It is likely that the only 'experiments' that will ever be done at such energies were completed about 10^{10} years ago when our Universe was less than one second old! It will not be easy to analyse the results, but we should try.

So, in this section we take a quick look at the Universe. There is a lot of it. With the naked eye it is possible (so I am told) to see about 7×10^3 stars. Already, Newton knew that their distances were at least

† A circular machine of $10^3\,M_p$ would have a diameter about 200 kilometres. That already looks hopeless—but there are other possibilities.

10^5 times the distance to the Sun. In fact, the nearest is 3×10^5 times the Sun's distance, i.e. about 1.5×10^{16} m from the Earth. Light from this star takes about $4\frac{1}{2}$ years to reach us.

The stars cluster together in 'galaxies'; a typical galaxy contains 10^{11} stars with an average separation of about 10^{17} m. The galaxies are shaped rather like discs, each having a radius around 3×10^{20} m and a thickness one-tenth of this. Intergalactic distances are about a hundred times their radii. If we reduce the scale by a factor $(3 \times 10^{20})^{-1}$ then galaxies will look like British pennies and in the Universe these pennies are about 1 metre apart. In fact the galaxies are not quite uniformly distributed, but tend to form clusters, which themselves group into superclusters. Within the range of the largest telescopes there are more than 10^{11} galaxies.

On the galactic (and smaller) scale the Universe is thus very 'lumpy'. However, on a larger scale it is in fact very smooth; that is, the average density of galaxies appears to be the same in whatever direction we look.

From the distribution of galaxies, and estimates of their masses, we can calculate an average matter density for the Universe:

$$\rho = 10^{-36}\,\mathrm{kg\,m^{-3}}. \tag{34.1}$$

By far the chief contributor to this density is hydrogen. The next most important element is helium with a mass abundance about a quarter that of hydrogen.

The Universe also contains many photons. Indeed, space is filled with randomly distributed photons with wavelength mainly in the range 0.1 to 1 cm. This is the so-called microwave background. The ratio of the number of protons N_p to the number of photons N_γ is given by

$$N_\mathrm{p}/N_\gamma = 10^{-9} \tag{34.2}$$

(with an 'error' of up to a factor 10). Because the photons have low energy their contribution to the energy density is, however, very small compared with that given by equation (34.1).

These photons provide the best evidence of the isotropy of the Universe, i.e. the fact that it is the same in all directions. The variation of this radiation falling on the Earth from mutually perpendicular directions is less than 1 part in 10^4.

Finally, we must know about the remarkable fact, discovered by

Hubble in 1929, though initially predicted, and rejected, by Einstein about ten years earlier (see §32), that the Universe is expanding. This expansion can be inferred from the fact that the light emitted from distant galaxies is 'red shifted'. To understand this, we recall that a given atom, in a hot gas, emits electromagnetic radiation (e.g. light) with a precise, fixed, set of frequencies. However, when the atom that emits the radiation is moving relative to the observer then the frequency is changed (the waves are either pushed together or stretched out according to the direction of the relative motion). For sources that are moving away from the observer the frequency is decreased, hence visible light is moved towards the red end of the spectrum.

It is easy to calculate the velocity from the observed shift in frequency and, in this way, Hubble discovered that any galaxy at a distance from the Earth measured by $\boldsymbol{R}$ was moving away from the Earth with a velocity $\boldsymbol{V}$ given by the universal rule:

$$\boldsymbol{V}=H\boldsymbol{R} \tag{34.3}$$

i.e. galaxies recede from us ($H>0$) at a velocity proportional to their distances. Note that this does not imply that everything is moving away from a 'centre' of the Universe (us); rather that *all* galaxies see all other galaxies moving relative to them† according to equation (34.3). A simple two-dimensional analogy is helpful here. We imagine the galaxies to be fixed marks on the surface of a balloon which is being inflated. The rule in equation (34.3) would then hold for all galaxies, regardless of which one was used as 'observer'.

The actual value of H (Hubble's constant) is somewhat uncertain, due to the difficulty of measuring distances, but

$$H^{-1}=(1\text{–}2)\times 10^{10} \text{ years} \tag{34.4}$$

probably covers the range of acceptable values. Note that this refers to present day observations. In spite of its name, H is unlikely to *be* constant in time; if we had observed at earlier times we would have found a larger value.

† Indeed the argument can be made in the other direction: if all observers are to see the same expansion, regardless of their position in the Universe, then the expansion *must* satisfy equation (34.3). In his book on the early Universe, *The First Three Minutes* (London: Deutsch 1977), Weinberg suggests that this fact may have helped Hubble to see this law in the primitive and inconclusive data which were available at the time.

An important consequence of equation (34.3) is that there is a limit to the observable Universe, a 'horizon' beyond which we cannot see. The origin of this can be realised if we note that, at some distance, equation (34.3) would imply relative velocities greater than c which, according to special relativity, should not be observable. It is this horizon, not the cost of big telescopes, which sets the ultimate limit on the size of the observable Universe.

Here we conclude our brief outline of some of the relevant properties of our Universe. In the next section we shall introduce the 'standard model' of cosmology† which describes the evolution of the Universe and enables us to understand some of these properties.

† Both particle theory and cosmology have a 'standard model'.

§ 35 The early Universe

The galaxies in the Universe are moving apart with velocities given by equation (34.3). It is an inevitable effect of gravity that this motion is slowing down, i.e. the relative velocity of any two galaxies is a decreasing function of time. Hence, if we go backwards in time, galaxies approach each other at an ever increasing rate and, eventually, all matter in the Universe comes together. The time when all distances become zero is the time when the Universe began and, according to the standard cosmological model, which over the last twenty years has become generally accepted, it began with a hot big bang.

To calculate when this occurred, i.e. to find the 'age' of the Universe, we initially ignore the effect of gravity. Then the relative velocities of galaxies can be assumed to have remained constant at the values given by $V = HR$. Their present separations, R, will then be given by this velocity multiplied by the time since the big bang. We therefore obtain the equation

$$R = H \times R \times \text{Age} \tag{35.1}$$

from which, on cancelling R, we find that the age of the Universe is the inverse of H. (Note that, because this is independent of the value of R, it is immaterial which pair of galaxies we consider; as is necessary for the calculation to make sense.)

The effect of gravity is to speed up the galaxies as we go *backwards* in time, and therefore to bring the big bang nearer. Thus the above result becomes an upper limit:

$$\text{Age} < H^{-1} \tag{35.2}$$

or, from equation (34.4),

$$\text{Age} < (1\text{–}2) \times 10^{10} \text{ years.} \tag{35.3}$$

Because of uncertainty in our knowledge of the present energy density in the Universe, it is not possible to do a precise calculation of the age in terms of H. However, it is unlikely that the above upper limit differs from the true age by more than a factor of about two.

We must now compare this estimate with other evidence on the age of the Universe. The earth has existed for about 4.5×10^9 years, so we are clearly in the right sort of region. We should pause here to take some pleasure in the fact that 'ages' determined by viewing distant galaxies, and by studying rocks on the Earth's surface which contain decaying nuclei, agree so well, especially when we realise how different are the theoretical bases and the experimental techniques involved. More recently it has become possible to measure the ages of the oldest star clusters in distant galaxies and figures of about 1.5×10^{10} years are found. This is already uncomfortably close to our limit in (35.2)†. Maybe there is the beginning of a real problem here, but, for the moment, we should again be encouraged that the two estimates are so close‡.

We shall now take a jump backwards in time so that we can look at the Universe when it is young. It is very hot and very dense; initially far hotter and denser than anything we can ever hope to observe again. It is expanding and cooling. The basic physics of such a system is reasonably well understood, but it is necessary to input knowledge of the elementary particles and their interactions. Matter, as we know it on Earth, will not exist under such conditions, since bound states will be broken apart by particle collisions. As the temperature falls, the collisions become less energetic (recall from §6 that the temperature is a measure of the average of the squared velocity, or the kinetic energy, of the particles) until, eventually, particular bound states can form and survive. When this happens the conditions can change dramatically and some of the properties of the world as we see it originate at such times.

For our first example of this we look at the Universe at an age of approximately 10^{12} s (3×10^4 years) when the temperature had fallen to about 3×10^3 °C. Previous to this the Universe had contained free

† And, incidentally, shows that the Earth has existed for only about a third of the life of the Universe.

‡ Readers may wonder how we measure the age of stars in distant galaxies, and may even doubt the accuracy of such measurements. They can begin their further study by reading the article 'Age of Universe Crisis Worsens', *New Scientist* **95** 829 (1982).

electrons and protons, but now these combined to form hydrogen atoms. The photons in the Universe, not having enough energy to excite these atoms, ceased to interact with matter. They became free to roam the Universe, and they are still with us. They are, indeed, the microwave background discussed in the previous section. Because of the expansion of the Universe they have been 'cooled', i.e. reduced in energy and hence increased in wavelength, an effect which can easily be calculated. They were born before our planet began, before the Universe contained matter, before the galaxies were created; yet their presence and their properties were correctly predicted prior to their being observed. The story of the microwave background is a great triumph for the big bang model.

For a second example, we look much earlier. By the time the Universe had reached an age of about 3 minutes, the temperature had dropped to 10^9 °C, so that particle energies became less than the binding energy of the deuteron. Thus all the protons and neutrons combined to form deuterons, which in turn joined to make helium nuclei. The process could not continue because two helium nuclei do not form a stable nucleus; also the electric repulsion between the charge 2 nuclei became an important inhibiting factor. This is the reason why heavy elements are comparatively rare. (Those that do exist were in fact created much later; they were 'cooked' in the interior of stars.)

But, we must ask, why did not *all* the protons and neutrons combine, so that there would be no hydrogen remaining? This would have been a disaster for the Universe, since helium is very uninteresting. Without hydrogen there would have been no life. The answer lies in the fact that there were not enough neutrons to go round; all the available neutrons *did* combine with protons but initially there were more protons than neutrons. Why? The reason for this lies in the small mass difference between neutrons and protons.

In order to understand the effect of the mass difference on the relative numbers we need to learn something from statistical mechanics. The probability of finding a particular particle (in a gas in equilibrium) depends on the ratio of its mass to the temperature. In fact, the ratio of neutrons to protons is given by

$$N_n/N_p = \exp[-(M_n - M_p)/T] \qquad (35.4)$$

where M_n and M_p are the neutron and proton masses, respectively, and T is the temperature, or actually, to make the units correct, the average particle kinetic energy. In this formula we have introduced the exponential function, denoted by exp, which will be well known to some readers. (Others will need to know that $\exp(x)$ is 1 when x is zero, is bigger than 1 when x is positive and smaller than 1 when x is negative). In order for some hydrogen (i.e. protons) to remain, we need N_n to be smaller than N_p, which, from equation (35.4), requires M_n to be larger than M_p. This, as we recall from the fact that the neutron is unstable, is the case. Indeed:

$$M_n - M_p = 1.3 \times 10^{-3} M_p. \tag{35.5}$$

The value of T which is required in equation (35.4) is the temperature below which deuterium is not being broken up by collisions. This corresponds to a temperature about equal to the deuteron binding energy which happens to be very similar to the value of $M_n - M_p$. In consequence, although the numbers of neutrons and protons are of the same order of magnitude, the ratio N_n/N_p is significantly less than unity, as required.

Detailed calculations, which involve considerations on the rate of the reactions and the rate of expansion of the Universe, give correctly the observed relative amounts of hydrogen and helium. We should be very impressed by this. We have calculated something that has only happened once, about 10^{10} years ago, when the Universe was very different from what it is now, and we have got the right answer! There will also be some deuterons remaining, which did not have time to combine into helium nuclei. It is possible to calculate the relative abundance of these, and again the agreement with observation is excellent.

It is interesting that the number of neutrino types enter these calculations (because they affect the rate of expansion, hence the precise temperature at which the deuterium begins to form and hence the ultimate fraction of helium which is made). If there were more than about 4 light neutrino types (i.e. more than four families) then the prediction of the hydrogen to helium ratio would be incorrect. This 'cosmological' limit on the number of light neutrinos is much smaller than the present 'particle physics' limit of about 20 (§29).

The above examples are well established successes of the big bang model (and of course of some aspects of particle theory). Our final example is more remarkable but is on a less secure basis. It concerns the ratio of protons to photons in the Universe (equation (34.2)) or, more generally, the question: why are there any protons at all? When the Universe was younger than about 10^{-6} s, it was hot enough for protons and antiprotons to be created in pairs, so that their numbers would be governed by statistical mechanics. This would imply that the number of protons and the number of antiprotons was approximately the same as the number of photons. When the temperature drops, creation would no longer be possible so the protons and antiprotons would collide and annihilate each other ($p+\bar{p}\rightarrow$photon(s) etc). They would *all* annihilate unless either: (i) there were more protons, say, than antiprotons initially, or (ii) they somehow became 'separated' in space.

If possibility (ii) occurred then we would expect that there would be some regions of space where, instead of matter being protons and electrons as on the Earth, it would be antiprotons and antielectrons. In this way we could still have the total of protons and antiprotons in the Universe equal, i.e. a zero total baryon number (like, apparently, the total electric charge). Although this is possible, we know that it does not happen in nearby regions of space, and it seems very unlikely. In particular, nobody has been able to think of any mechanism whereby protons and antiprotons could become spatially separated.

Thus we appear to be left with (i). If we then use the result in equation (34.2), together with the fact noted above that, prior to the time when annihilation dominates, proton and photon numbers are approximately equal, we reach the strange conclusion that when the Universe had cooled to about 10^{13} °C, there were $(1+10^9)$ protons for every 10^9 antiprotons. This rather looks as though God tried to make the baryon number zero but just missed. Of course the extra one in the 10^9 is the important portion. Without it the Universe would be all photons; there would be no 'matter'.

There is, however, a possibility of explaining this strange number, 1 in 10^9. To understand this we recall that in grand unified theories, baryon number is not conserved. It may well be that at the initial big bang the baryon number of the Universe was zero (which would imply $N_p=N_{\bar{p}}$ etc), but that at some later time excess baryon number was created. This could only have happened at a temperature where the baryon creation terms can operate, e.g. where the X particle of §31 can exist. This requires temperatures around 10^{27} °C which occurred when the Universe was 10^{-35} s old. What took place at that time may have been essential for there to be any matter at all in today's Universe! The advantage of this idea is that we do not have to input the strange 1 part in 10^9 proton/antiproton asymmetry; it becomes a calculable number.

Not surprisingly, the calculation is subject to many uncertainties. For example, since the effect we seek is clearly not time-symmetric (baryon number must *increase*, not *decrease*), it is important, for it to occur, that the 'gas' should not be in equilibrium, which means that it should be expanding faster than the rates of the various interactions. Also, of course, we need an interaction which is not symmetric between baryons and antibaryons. The only thing we can say from the present 'estimates' is that the mechanism *might* work.

Further back to the big bang we cannot go. At around 10^{-47} s the temperature was so high that particles had kinetic energies equal to the Planck mass, and effects of quantum gravity must have been important. Unfortunately, we do not have much idea how to calculate such effects.

The standard model of cosmology, like the standard model of particle theory, works; it fits in well with what we have learned about the Universe and about elementary particles. There are, however, some problems, which appear to require an amazing modification to the model. Fortunately, the ideas that we have developed earlier may provide the mechanism to drive this modification, as we shall see.

§ 36 The inflationary Universe

Although in many ways very satisfactory, the hot big bang model, as we have described it in the previous section, meets three serious problems: the 'horizon', 'flatness' and 'monopole' problems.

To explain the first of these we recall the isotropy of the Universe as demonstrated, for example, by the microwave background. In general, all parts of the observable Universe appear to be the same. This may not seem surprising, until we realise that the amount of material in the observable Universe grows with time. This is because relative velocities decrease with time due to gravity, hence galaxies that initially were receding with velocities greater than the velocity of light, i.e. were outside the horizon and so unobservable, can slow sufficiently so as to have velocities less than c, thereby becoming observable. Now comes the problem: how does a galaxy, newly entering our observable Universe, 'know' what we are like so that it can be the same? *It has never ever in its previous history been able to 'communicate' with anything inside our horizon.* We try to illustrate this in figure 36.1. In this figure d_{now} is the present diameter of the observable Universe, i.e. of the horizon. At an earlier time (t_1) this will have shrunk, because of the Hubble expansion, to a diameter $d_{\text{now}}(t_1)$. However, the horizon will shrink much faster so that its diameter at t_1, i.e. d_1, will be smaller than $d_{\text{now}}(t_1)$. Thus, at t_1, the three Universes A, B and C will be entirely separate; they will never have been able to communicate. Who arranged that they should have the same density? Indeed why should they have started at the same time?† It is as if people from different galaxies, who had never had direct or indirect communication used, for example, the same symbols for the integers. To appreciate the magnitude of the problem it is worth noting that our present observable Universe would have separated into more than 10^{80} such separate Universes by the time the temperature reached the GUT mass scale. This is the 'horizon' problem.

† Spinoza (see footnote on page 157) would have enjoyed this problem!

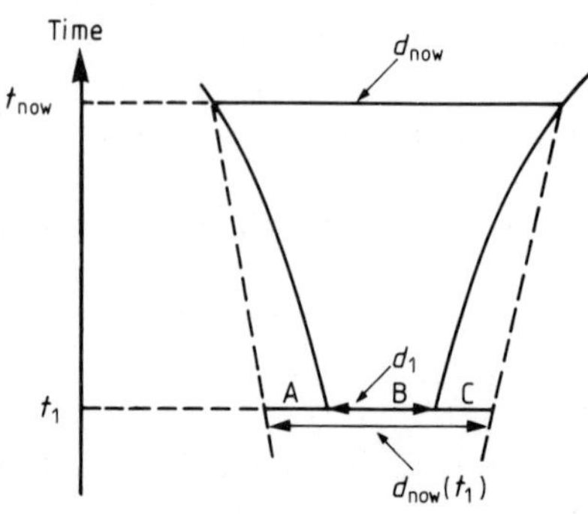

Figure 36.1 Illustrating the 'horizon' problem. Parts of the Universe which are now (at t_{now}) inside the observable Universe were once (e.g. at t_1) outside it. How could they 'know' about each other?

For the second problem we consider the average density of the Universe. There exists a 'critical density' given by

$$\rho_{crit} = 3H^2/8\pi G \tag{36.1}$$

such that, if the actual density is above this, then gravity will slow down the rate of expansion until it stops. It will then reverse and the Universe will recollapse to a point (the 'big crunch'). Alternatively, if the actual density is below this, then it will expand forever.

It would be interesting to know the real situation, but it turns out that the errors are such that we cannot be sure. All we can say is that we are reasonably close to the critical density:

$$10\rho_{crit} > \rho > 10^{-1}\rho_{crit}. \tag{36.2}$$

This is already rather remarkable. In our story we have met many very large, and very small numbers—why should the real density be so close to the critical density? It could, of course, be a coincidence.

On closer examination, however, the coincidence becomes even more remarkable. Consider the Universe at a very early time. It will have some density which, in general, will not be the critical density appropriate to its rate of expansion, so it will either collapse or expand for ever. Now we can calculate the timescale of the collapse, for example, i.e. the 'lifetime' of the Universe. It will be formed from the only constants available, namely G, c and h. These give the Planck time, which is 5×10^{-44} s! (See §32.) This is the lifetime of a typical collapsing Universe. On the other hand, an expanding Universe will normally expand so fast that in the same sort of timescale its density will become negligible; it will have dissipated into separate particles. Why, then, has our Universe already 'lived' for 10^{10} years ($=3\times10^{17}$ s), which is about 10^{61} times the expected life? For this to have happened it is necessary for the initial density to be fantastically close to the initial critical density. In fact, detailed calculations show that at an early time in the Universe the density was equal to the critical density to an accuracy of 1 part in 10^{55}! Here we have yet another fine-tuning problem; because of the similarity of the metric when $\rho=\rho_{\text{crit}}$ to the metric of flat space, it is called the 'flatness' problem.

To describe the third problem we first mention a strange asymmetry in electromagnetism. Electric and magnetic fields are very similar, but the former are related to static electric charges and the latter to moving charges. It is possible to introduce 'magnetic charges' which are related to the fields in the opposite way. Such objects, and none have been seen, are called 'magnetic monopoles'. (A magnet has a 'north' and 'south' pole—if we could separate one of these it would be a monopole. But we can't.) Now gauge theories, with the symmetry broken by a Higgs mechanism, tend to create monopoles. Indeed, we can calculate the density of such monopoles in the standard model, and it turns out to be large. In fact, monopoles would be the major contribution to the average density in the Universe, and would give a density vastly too large to be consistent with other evidence. This is the 'monopole' problem.

These three problems can probably be solved in the framework of the 'inflationary Universe' proposal of Kazanas and Guth. This uses the cosmological constant mentioned in §32. Although the present value of this is small, compared for example to the average density of the Universe, it could have been large once and then been cancelled by a contribution to the energy density caused by a symmetry-breaking mechanism. Recall that, in such a mechanism, a field variable drops into a lower energy state. It is possible to imagine that this process takes a long time and that during this time the Universe is in a type of 'supercooled' state. The energy available can then drive a rapid expansion. It could well be that *all of our present Universe* was contained inside a small 'bubble' at a time around 10^{-35} s, and that this bubble then inflated (growth by factors up to 10^{3000} have been proposed!). Such an inflation would appear to solve the three problems above. There are many details into which we cannot go; for example, we do not want to make the resulting Universe *too* homogeneous, otherwise it would have no galaxies; but the present situation seems promising. Unfortunately, recent developments suggest that we need a *separate* symmetry-breaking mechanism just to drive inflation; those that are introduced for purely particle physics reasons do not work. However the whole issue is very much part of today's research; it is a study where cosmology and particle theory become essentially linked; it may be that its solution will uniquely merge gravity with particle physics; it is a good place to close this section and—almost—our story.

§ 37 Coda

All day long the trains run on rails. Eclipses are predictable. Penicillin cures pneumonia and the atom splits to order. All day long, year in, year out, the daylight explanation drives back the mystery and reveals a reality usable, understandable and detached. The scalpel and the microscope fail, the oscilloscope moves closer to behaviour. The gorgeous dance is self-contained, then; does not need the music which in my mad moments I have heard. Nick's universe is real. All day long action is weighed in the balance and found not opportune nor fortunate or ill-advised, but good or evil. For this mode which we must call the spirit breathes through the universe and does not touch it; touches only the dark things, held prisoner, incommunicado, touches, judges, sentences and passes on.
Her world was real, both worlds are real. There is no bridge†.

In the preceding sections we have tried to explain, and even to enjoy, the success that physicists have had in exploring our world. Understanding so much, there is a danger that we might believe we understand almost everything. However, as is normally the case, those who claim to 'know all the answers' are merely ignorant of the questions, so in this final section we make brief reference to a few of the areas where we (or, to be strictly accurate, 'I') understand very little.

We begin by thinking about time. Encouraged by the theory of relativity, we have tended to regard the time variable as rather like a space variable. Apart from the occasional change of sign, time and space enter the equations of physics in a very similar way. However, we all *know* that time and space are very different. In case our story has almost convinced us otherwise we should consider a few examples. I am *here* (a point in space); I could of course be *there* (another); or even *there'* (etc). Indeed, I can go to there, and then come back to here. Try it—it really works! I am also at *now* (a point in time), but I cannot stay at now, and I have no idea how to go to *then*

† W Golding, *Free Fall* (London: Faber 1959).

(another point in time) or to return from then to now. The idea we are trying to convey is that time 'flows'. Although it is probably not completely clear what this means, it is certainly true, and it is a fact of which physics is totally unaware. Within physics there is nothing which corresponds to a 'now'.

A (perhaps) related phenomenon concerns direction. From where I now sit I could go to the left or to the right with equal facility; apart from the odd wall, etc, the positive x direction is not of a different nature from the negative x direction. With time, however, there is an evident *qualitative* difference. I remember yesterday, but I *cannot* remember tomorrow! We experience a world that has a manifest lack of time-reversal symmetry. This familiar, and apparently trivial, fact becomes very significant when we realise that all the basic laws of physics are invariant under the change $t \rightarrow -t$. Fundamental physics cannot distinguish the past from the future, so where should we look to find the source of the manifest asymmetry?

Some readers may wish to query the last two sentences, and, certainly, they should be qualified. It is true that there are very weak effects, so far seen only in decays of the K meson, which do violate time-reversal symmetry. It is, however, surely inconceivable that such effects are relevant to this discussion, at least as far as they affect what happens at present. It may be that at some earlier stage of the Universe they were responsible for effects that are important now, e.g. generation of the baryon excess as discussed in §35.

The other qualification concerns things like the second law of thermodynamics which, roughly speaking, tells us that systems 'run down', i.e. they change from being highly ordered, or arranged, to being more smooth and random. As an example, suppose we bring a hot body into contact with a cold one; then heat will flow so as to bring them both to the same temperature. Initially the molecular energies in the hot body were, on average, higher than those in the cold body, but in time these energies become randomly distributed between the two bodies, i.e. both reach the same temperature. It is clear that this effect requires a direction of time, hence even physics knows about time asymmetry. However, this is not a part of *fundamental* physics. The second law of thermodynamics is sometimes claimed to be *derived* from the fundamental laws. Indeed many such 'derivations' exist. They are all 'false', in that they use only laws that do not have a time-reversal asymmetry. In fact a time direction is *input* to these derivations. It is an indication of how totally prejudiced we are about the direction of time that we do not readily see where the crucial asymmetry assumption is inserted in these 'proofs'. Of course the prejudice can readily be understood: the second law of thermodynamics does work; almost everything we see around us is directed in time—plants grow, wither and decay, cars are made and then rust, runners become hot and tired, the number of romantic novels in the world increases—we do not observe these processes being reversed.

A good way to see what is happening here is to consider a film of a man shuffling a pack of cards. Before and after each shuffle he lays them out so that we can see the order. Everything is here reversible in time and we could not detect any difference if the film were run backwards. However if he had begun with a prepared pack of cards, in the order: ace to king of hearts, etc, then we *would* know which was the correct direction for the film. It would be inconceivable that a random shuffle would produce perfect ordering†. In the same way, the second law of thermodynamics works for large systems because there is an enormous amount of order in the world, so random processes almost invariably reduce it. We observe the world to be 'running down', because it is at present so enormously 'wound up'. But, *how did it become wound up in the first place*? Was there some time in the Universe when everything was working the other way?

The answer to these questions has probably got something to do with the big bang, which does appear to provide a time direction (though not if at some future time there *is* a big crunch). Notice that in our discussion on the early Universe we *assumed* that the Universe is determined by the initial conditions rather than by those at a later time—maybe this is too large an extrapolation of current prejudices!

This is a fun problem to think about‡, but we shall allow it to lead us to something different. Time is the medium of change, and physics is very much concerned with the way things change. Moreover our story has, I hope, revealed some of the remarkable changes of attitude that have been forced upon physicists and that have allowed progress to be made. The success of our story has depended upon a willingness

† Of course, this sequence has exactly the same probability as any other—i.e. about 1 in 10^{68}—it is just that there are a lot of 'others'.

‡ In summarising a conference on time held in 1967, Wheeler referred to a 'feeling of not really understanding anything' (*The Nature of Time* ed Gold (Ithaca: Cornell University Press 1967) p 233).

to abandon old ways of thinking and accept new ones. This same willingness is often missing in the world outside physics. Perhaps the most striking, and certainly the most potentially dangerous, aspect of this is the unwillingness of the world's politicians, and, to a large extent, of the media which are their mouthpiece, to imagine a world in which international disputes are solved other than by a resort to violence. There is no necessity for the use of violence among nations; weapons, 'defence systems', the whole sordid business of militarism and the childish petulance of the language of politicians, are as irrelevant to the real problems of humanity as are mechanical models of the ether to QCD. Only a massive inertia and unwillingness to change prevents the resources of the world being used to remove hunger and disease rather than being squandered on bombs. The 'argument' that wars have always happened and therefore always will, is a good example of this inflexibility of attitude. 'History' is the story of *change*; its only lesson is that it has no lessons. Once, men died of diseases which now are cured, once they had to draw their water from wells, once they lit their houses with candles and had to do sums with 'long division', etc. Maybe, one day, they will look back and say, with a smile, 'once people made bombs to throw at each other'.

All this may not have much to do with our story, but physicists are guilty of having given people the power to destroy themselves, so it is time we did something to discourage this folly. Also, it is possible—not, perhaps, likely, but certainly possible—that we are alone in the Universe, that we are the *unique*, *conscious*, *observers* of the whole marvellous show†. It would then be a disaster of cosmic proportions if we were to destroy ourselves!

This, naturally, brings us to our next topic. It is important that 'physics' should know that *we are here*. Assuming that physics had no particular preferences either way on the issue of our existence, that fact is an almost incredible 'accident'. If we were to make tiny changes in the parameters of physics, at all levels: the neutron–proton mass difference, nuclear masses, atomic energy levels, initial big bang densities, etc, then we would not, *could not*, exist. Note that it is not that we would be a little different; rather that there would be no possibility of complex structures—still less of life—in any form at all!

We saw an example of this already in §35, where we noted that the

† If there are others then it is puzzling why we have not, it seems, seen any sign of them—see, for example, F J Tipler's article in *New Scientist* **96** 33 (1982).

mass difference between the proton and neutron was crucial to the survival of hydrogen, without which there could be no life. Conversely, it would require only a small reduction in the strength of nuclear forces to make the deuterium nucleus unbound; then, no deuterium, no helium, hence no heavy elements, leading to a Universe of hydrogen—vast, evolving and *unseen*! An equally small increase in the nuclear force would have different, but equally disastrous, effects. Our existence appears to be dependent on many accidents†. Can we understand this?

The solution is usually given in terms of the 'anthropic principle', which has various forms, all of which depend on the rather obvious fact that we can only observe a world in which it is possible for us to exist. Thus the 'accidents' *have* to happen. One can understand this if there are many Universes, all with different physics; most will be totally uninteresting and will not contain anything remotely like life. One possible source of many Universes would be if the big bang were really a 'big bounce'—the Universe expands and then contracts to some sort of singularity at which all physics changes, then the cycle repeats itself. Eventually it is possible that we might appear, though it seems unlikely that this would happen before the Universe went into a situation where it did not recontract but simply went on expanding for ever. An alternative possibility is suggested by the inflating bubbles model of the previous section. Maybe the bubbles exist in an infinite 'bath', in which case any event, however impossible, would happen (indeed it would happen an infinite number of times). Of course, if something like this is the answer then it would appear that our attempts to understand the world in terms of simple theories containing few parameters are misguided. If there is really little 'freedom' in making physics, it would surely be unbelieveable that within the permitted freedom one could make a world that could produce us. We should not expect our world to be 'simple', rather to be as simple as possible, consistent with *our* existence‡.

Before we leave the anthropic principle we note that it is, in some vague sense, a continuation of the developments in quantum theory and relativity. In both these theories it was necessary to abandon

† See *The Accidental Universe* by P C W Davies (Cambridge: Cambridge University Press 1982) for further discussion.

‡ For some development of this idea see the author's article 'Do we live in the simplest possible interesting world?', *Eur. J. Phys.* **2** (1981) 55.

previous notions of absoluteness and to elevate the importance of the *observer*. Maybe observers are not only necessary to understand particular phenomena—they may be essential for there to be a Universe at all.

This is clearly a big and fascinating subject. Physics has to permit the existence of physicists. Naturally the idea of 'design' enters here. If our existence is coming to look like an amazing accident, then maybe what we are learning is that we are in some sense the object of the Universe. This is perhaps not different from saying that what happens is at least partially determined by the future, as well as (or instead of) the past. We return to this topic shortly but here we should note that, presumably, even a designer had to operate within some laws. Whether it was 'chance' or 'God' that selected a world that could produce people, it is still amazing. In either case, there could not have been many viable options. Rather than protest that God should have made a better world, we should rejoice that he found it possible to construct one at all—only in hymns is God omnipotent!

Let us return to our story. We have learned much by taking objects apart, by looking at constituents. But we must remember that in this process of finding out more, we lose something. When we look at the individual words in a play, we have lost the plot; when we look at the atoms in a pencil, we cannot write; when we look at the protons and neutrons in a nucleus, the atom has lost its identity; etc. The constituents do not contain everything—the way they are put together is important also; many properties are a consequence of 'organisation'. I am prompted to mention this because I recently heard a microbiologist say that frogs were very like people (I think because they contained similar proteins, etc). I have no reason to believe that the statement, in its strict context, was other than correct, but he went on to mock 'thinkers of the Middle Ages' who thought that frogs and people were different. At this stage my sympathies were with the Middle-Age thinkers. Whatever we may mean by 'difference' (and, of course, as we recall from §32, we need a metric to measure a difference), on any reasonable general definition, people and frogs are 'very different'. What he really meant was that this difference (noticed not only in the Middle Ages) was not seen in his particular method of observation, which was perfectly adequate for his primary purpose, but not for the more general purpose. Of course we can now see the point very clearly; frogs and people, or men and women, etc, are all made of identical quarks and leptons. This fact does not make them the same—organisation is crucial!

Our next topic concerns quantum theory. In formulating the rules of quantum theory it is usual to introduce the idea of an 'observer'. An object is a particle or a wave according to how it is observed; a wave-function gives the probability for a particle to be in a particular place, say, when its position is measured. The question naturally arises as to what constitutes an 'observation' or a 'measurement', and the only answer we can give is that it must involve something which is outside the framework of *quantum theory*. As long as the measuring apparatus (or 'observer') is described by quantum theory, then it is part of the wave function of the system. This function will satisfy a particular equation which implies that its change with time is precise and predictable; no 'probability' features will enter. These can only be introduced from outside through *non-quantum* observation. This difficulty with the basic structure of quantum theory was realised in its early days (in particular, by two of the leading contributors to the theory, Einstein and Schrödinger) and has been the subject of debate, among physicists and philosophers, ever since. Recently, modern experimental techniques have permitted accurate experimental tests of some of the predictions of quantum theory which appear to contradict 'common sense', and some very beautiful verifications of the theory have been obtained. Naturally this has aroused renewed interest in the interpretation problem which lies at the heart of the theory. It remains, however, an unsolved problem. Quantum theory is very elegant, it is apparently always correct; unfortunately, however, the closer we enquire into what it means, the more puzzling it becomes.

It could be that the answer lies with gravity; it could be that there is some clue to be found by thinking of the connection between the direction of time and the basic time-irreversibility of the process of measurement; it could be that the enormous number of molecules that are required in any realistic measuring device is somewhat crucial; most intriguing of all is the possibility that 'observation' requires *conscious* observers, that we can connect this most mysterious and obscure aspect of physics with the most mysterious and obscure property of existence, namely, '*consciousness*'.

As a theoretical physicist I, along with all other physicists, know that I operate (like all other pieces of apparatus) according to well defined laws, I know that the particles within me interact in specific ways and determine my behaviour; I know also that I am an utterly insignificant part of the Universe which goes along its mechanistic path neither knowing nor caring about me. But this 'knowledge' has little effect on me—I, along with every other individual, am the centre of my Universe, galaxies can form or collapse but they concern me far less than personal issues which involve times and spaces that are negligible on galactic scales; I struggle to 'choose' the form that my (predetermined!) actions will take, I spend time making decisions, etc. These are of course very familiar facts to all of us but, in the context of our story, they are very remarkable—indeed, they are (by many orders of magnitude) the most remarkable facts of our world. They are symptoms of the fact that I am *conscious* and I am *free.* (Incidentally, it is an empty question whether free will is an 'illusion'. Freedom is a property of the conscious mind: if I *think* I am free then I *am* free.) But what is consciousness? And where within me is it to be found? Is it in the quarks and gluons? Is it in the way they are organised—something that occurs whenever there is a sufficient degree of complexity? Or is it, somehow, separate and outside of these things?

I do not know what consciousness is, i.e. I have no idea how to define it in other terms. That it exists is not in dispute. It is, to each of us, our own vast Universe, private and personal, in which we apparently reign supreme. It gives rise to what is surely the cleverest trick of evolution—*purpose.* Outside my window there are motor cars. To explain these structures in terms of the motion of elementary particles, following precise laws since the early Universe, would be a difficult and probably foolish exercise (though—at least in so far as the Universe is a causal system—its initial condition determined the existence of the Renault 14); we see them more naturally as the outcome of design and purpose. As we noted earlier, an alternative way of expressing this, is to say that what happens is more readily understood in terms of the *final* state (the completed car) than the *initial* state. Maybe the direction of time, the measurement problem in quantum theory and the understanding of consciousness are all somehow related.

It is a natural, but by no means universally accepted, step, from the clear evidence of design and purpose we see in our immediate environment, to the belief that there is design and purpose in the origin of the Universe. I personally am willing to take that step: the physical Universe did not create consciousness but, in a way that I do not comprehend, was created by it. Physics exists through a design and purpose in the conscious mind of the one whom I call God. Some of my readers, with a different cultural or linguistic background, will, rightly, use a different name. Others will not follow me into what they will see as unnecessary speculation.

In the book of Genesis we are told that God, beholding all that he had made, was pleased. However much, or little, that statement may mean to us, we can at least share the pleasure as we, too, 'acknowledge the wonder' of all that is, thankful that we are privileged to be part of it.

Appendix on units

In this appendix we shall endeavour to clarify a few technical points concerning units. First, we must distinguish carefully between two aspects of the determination of the unit to be used for a given quantity. One, which is by far the most significant physically, concerns the *nature* of the quantity. Some of these are *basic*, e.g. length, time, mass. Others are *derived*, e.g. volume, which is $(\text{length})^3$, mass-density which is mass per unit volume, i.e. $(\text{mass}) \times (\text{length})^{-3}$, etc. Of course, the nature of a quantity is not a matter of choice, it is determined by the physics. On the other hand there is some degree of choice involved in deciding which quantities to regard as basic; we could, for example, replace length by velocity as a basic unit, in which case length would be the derived quantity (velocity) × (time). There is a more subtle type of choice involved in deciding *how many* quantities to regard as basic. We give an example of this below.

The aspect of units to which we are referring here is usually known as the 'dimension' of a quantity. Thus, for example, we say that density has the dimensions of $(\text{mass}) \times (\text{length})^{-3}$. We do not, in this book, use the word dimension in this context, however, because of possible confusion with the other use of the word when we speak, for example, of space as having three dimensions (all of which are of course *lengths*).

The second aspect of units is entirely a matter of choice, and concerns the size we take as 1 unit. Do we measure a length in miles, inches, centimetres or whatever? Clearly the most appropriate choice depends on the 'scale' of the objects being considered. There is, however, an internationally agreed set of units (SI units) which are based on the metre (m), kilogramme (kg) and second (s). In most cases we have used these units in the text, though, fortunately, it is still considered acceptable not to express dates as so many seconds AD.

We now consider the choice of units in electromagnetism. By keeping the parameters α and β of §§7 and 10 arbitrary, we have avoided the need to make such a choice. For completeness, however, we shall discuss here two possibilities.

The simplest of these is to omit the factor α in §7 (or, equivalently, put it equal to one). Then Coulomb's law

$$\boldsymbol{F}=\frac{Q_1Q_2}{r^2}\hat{\boldsymbol{r}} \tag{A1}$$

itself defines the unit of charge and allows it to be expressed in terms of other units:

$$(\text{charge})=(\text{mass})^{1/2}(\text{length})^{3/2}(\text{time})^{-1}. \tag{A2}$$

Thus, with this procedure, no new basic charge unit is required. The equation giving the magnetic field becomes (using equation (10.4))

$$\boldsymbol{B}=\frac{Q}{c^2}\frac{\boldsymbol{v}\wedge\hat{\boldsymbol{r}}}{r^2}. \tag{A3}$$

In spite of the formal simplicity of the above, the SI scheme, with which most readers will be familiar, proceeds differently. One reason for this is that it is easier to measure the force between two currents than between two charges. We therefore introduce a unit of charge called the *coulomb*, defined by the statement that the force per metre between two long straight wires 1 metre apart, each carrying a current of 1 coulomb per second, is 2×10^{-7} kg s^{-1}.

This force can easily be calculated from equations (10.2) and (10.3) and we find that the value required for β becomes 10^{-7} C^{-2} kg m. In fact, it is usual to write equation (10.3) in the form

$$\boldsymbol{B}=\frac{\mu_0}{4\pi}Q\frac{\boldsymbol{v}\wedge\hat{\boldsymbol{r}}}{r^2} \tag{A4}$$

with

$$\mu_0/4\pi=10^{-7}\,\text{C}^{-2}\,\text{kg m}. \tag{A5}$$

Similarly, when using SI units it is usual to write Coulomb's law (equation (7.3)) in the form

$$\boldsymbol{F}=\frac{1}{4\pi\varepsilon_0}\frac{Q_1Q_2}{r^2}\hat{\boldsymbol{r}} \tag{A6}$$

where

$$4\pi\varepsilon_0c^2=10^7\,\text{C}^2\,\text{kg}^{-1}\,\text{m}^{-1} \tag{A7}$$

or (approximately)

$$\varepsilon_0 = (10^{-9}/36\pi)\ \mathrm{C}^2\,\mathrm{kg}^{-1}\,\mathrm{m}^{-3}\,\mathrm{s}^2. \qquad \text{(A8)}$$

There is of course a natural, fundamental, unit of charge, namely, the charge on an electron. All observed charges are integral multiples of this charge. Incidentally, this fact itself is something of a puzzle. Why should the proton charge be, apparently, *exactly* equal to minus the electron charge? Similarly, why should the *u*, *d* quarks have charges which are exactly $-\frac{2}{3}$, $+\frac{1}{3}$ of the electron charge? The composite model of §28 offers a possible explanation—but this depends on the charge on the V rishon being exactly zero, which is perhaps equally hard to explain.

Some suggestions for further reading

There are several popular books available covering essentially the story we have told here, generally with fewer details and less mathematics. I like B K Ridley's *Time, Space and Things* (2nd edition, Cambridge 1984), although *Quarks: The Stuff of Matter* (H Fritzsch, Allen Lane/Basic Books 1983) and *The Cosmic Onion: Quarks and the Nature of the Universe* (F Close, Heinemann 1983) give more emphasis to recent developments. A more detailed account of many of the experiments to which we have referred briefly is given by S Weinberg in *The Discovery of Subatomic Particles* (Scientific American Library 1984). H Pagels' *The Cosmic Code* (Simon and Schuster 1982) is very readable and contains a good discussion of the interpretation problem of quantum theory, as does the recent book of J C Polkinghorne, *The Quantum World* (Longman 1983), and, even more thoroughly, the delightful book *In Search of Reality* by B d'Espagnat (Springer 1983).

The Left-Hand of Creation by J D Barrow and J Silk (Heinemann 1983) and *The First Three Minutes* by S Weinberg can be recommended for more about cosmology and the early Universe. A good account of the many apparent accidents which are prerequisites for life in the Universe, and of the anthropic principle, is given by P C W Davies in *The Accidental Universe* (Cambridge 1982). The best book on the general philosophy of physics which I have found is that by G Toraldo di Francia, *The Investigation of the Physical World* (Cambridge 1981; the original Italian version *L'Indiagine de/Mando Fisicer* was published in 1976).

Finally, readers who wish to study particle physics more seriously should obtain *Quarks and Leptons* by Halzen and Martin (Wiley 1984).

Index

Page numbers in bold type refer to complete sections; suffix n refers to footnotes.